W9-CCT-685

# ECOLOGY

## *Earth's Living Resources*

**Anthea Maton**
Former NSTA National Coordinator
Project Scope, Sequence, Coordination
Washington, DC

**Jean Hopkins**
Science Instructor and Department Chairperson
John H. Wood Middle School
San Antonio, Texas

**Susan Johnson**
Professor of Biology
Ball State University
Muncie, Indiana

**David LaHart**
Senior Instructor
Florida Solar Energy Center
Cape Canaveral, Florida

**Maryanna Quon Warner**
Science Instructor
Del Dios Middle School
Escondido, California

**Jill D. Wright**
Professor of Science Education
Director of International Field Programs
University of Pittsburgh
Pittsburgh, Pennsylvania

Prentice Hall
Englewood Cliffs, New Jersey
Needham, Massachusetts

*Prentice Hall Science*

# Ecology: Earth's Living Resources

**Student Text and Annotated Teacher's Edition**
**Laboratory Manual**
**Teacher's Resource Package**
**Teacher's Desk Reference**
**Computer Test Bank**
**Teaching Transparencies**
**Product Testing Activities**
**Computer Courseware**
**Video and Interactive Video**

The illustration on the cover, rendered by Joseph Cellini, shows humpbacked whales, one of many whale species that are endangered.

Credits begin on page 160.

SECOND EDITION



ISBN 0-13-225558-8

3 4 5 6 7 8 9 10 97 96 95 94

Prentice Hall
A Division of Simon & Schuster
Englewood Cliffs, New Jersey 07632

## STAFF CREDITS

**Editorial:** Harry Bakalian, Pamela E. Hirschfeld, Maureen Grassi, Robert P. Letendre, Elisa Mui Eiger, Lorraine Smith-Phelan, Christine A. Caputo
**Design:** AnnMarie Roselli, Carmela Pereira, Susan Walrath, Leslie Osher, Art Soares
**Production:** Suse F. Bell, Joan McCulley, Elizabeth Torjussen, Christina Burghard
**Photo Research:** Libby Forsyth, Emily Rose, Martha Conway
**Publishing Technology:** Andrew Grey Bommarito, Deborah Jones, Monduane Harris, Michael Colucci, Gregory Myers, Cleasta Wilburn
**Marketing:** Andrew Socha, Victoria Willows
**Pre-Press Production:** Laura Sanderson, Kathryn Dix, Denise Herckenrath
**Manufacturing:** Rhett Conklin, Gertrude Szyferblatt

## Consultants

Kathy French — National Science Consultant
Jeannie Dennard — National Science Consultant
Brenda Underwood — National Science Consultant
Janelle Conarton — National Science Consultant

## Contributing Writers

Linda Densman
*Science Instructor*
*Hurst, TX*

Linda Grant
*Former Science Instructor*
*Weatherford, TX*

Heather Hirschfeld
*Science Writer*
*Durham, NC*

Marcia Mungenast
*Science Writer*
*Upper Montclair, NJ*

Michael Ross
*Science Writer*
*New York City, NY*

## Content Reviewers

Dan Anthony
*Science Mentor*
*Rialto, CA*

John Barrow
*Science Instructor*
*Pomona, CA*

Leslie Bettencourt
*Science Instructor*
*Harrisville, RI*

Carol Bishop
*Science Instructor*
*Palm Desert, CA*

Dan Bohan
*Science Instructor*
*Palm Desert, CA*

Steve M. Carlson
*Science Instructor*
*Milwaukie, OR*

Larry Flammer
*Science Instructor*
*San Jose, CA*

Steve Ferguson
*Science Instructor*
*Lee's Summit, MO*

Robin Lee Harris Freedman
*Science Instructor*
*Fort Bragg, CA*

Edith H. Gladden
*Former Science Instructor*
*Philadelphia, PA*

Vernita Marie Graves
*Science Instructor*
*Tenafly, NJ*

Jack Grube
*Science Instructor*
*San Jose, CA*

Emiel Hamberlin
*Science Instructor*
*Chicago, IL*

Dwight Kertzman
*Science Instructor*
*Tulsa, OK*

Judy Kirschbaum
*Science/Computer Instructor*
*Tenafly, NJ*

Kenneth L. Krause
*Science Instructor*
*Milwaukie, OR*

Ernest W. Kuehl, Jr.
*Science Instructor*
*Bayside, NY*

Mary Grace Lopez
*Science Instructor*
*Corpus Christi, TX*

Warren Maggard
*Science Instructor*
*PeWee Valley, KY*

Della M. McCaughan
*Science Instructor*
*Biloxi, MS*

Stanley J. Mulak
*Former Science Instructor*
*Jensen Beach, FL*

Richard Myers
*Science Instructor*
*Portland, OR*

Carol Nathanson
*Science Mentor*
*Riverside, CA*

Sylvia Neivert
*Former Science Instructor*
*San Diego, CA*

Jarvis VNC Pahl
*Science Instructor*
*Rialto, CA*

Arlene Sackman
*Science Instructor*
*Tulare, CA*

Christine Schumacher
*Science Instructor*
*Pikesville, MD*

Suzanne Steinke
*Science Instructor*
*Towson, MD*

Len Svinth
*Science Instructor/*
*Chairperson*
*Petaluma, CA*

Elaine M. Tadros
*Science Instructor*
*Palm Desert, CA*

Joyce K. Walsh
*Science Instructor*
*Midlothian, VA*

Steve Weinberg
*Science Instructor*
*West Hartford, CT*

Charlene West, PhD
*Director of Curriculum*
*Rialto, CA*

John Westwater
*Science Instructor*
*Medford, MA*

Glenna Wilkoff
*Science Instructor*
*Chesterfield, OH*

Edee Norman Wiziecki
*Science Instructor*
*Urbana, IL*

## Teacher Advisory Panel

Beverly Brown
*Science Instructor*
*Livonia, MI*

James Burg
*Science Instructor*
*Cincinnati, OH*

Karen M. Cannon
*Science Instructor*
*San Diego, CA*

John Eby
*Science Instructor*
*Richmond, CA*

Elsie M. Jones
*Science Instructor*
*Marietta, GA*

Michael Pierre McKereghan
*Science Instructor*
*Denver, CO*

Donald C. Pace, Sr.
*Science Instructor*
*Reisterstown, MD*

Carlos Francisco Sainz
*Science Instructor*
*National City, CA*

William Reed
*Science Instructor*
*Indianapolis, IN*

## Multicultural Consultant

Steven J. Rakow
*Associate Professor*
*University of Houston—Clear Lake*
*Houston, TX*

## English as a Second Language (ESL) Consultants

Jaime Morales
*Bilingual Coordinator*
*Huntington Park, CA*

Pat Hollis Smith
*Former ESL Instructor*
*Beaumont, TX*

## Reading Consultant

Larry Swinburne
*Director*
*Swinburne Readability Laboratory*

# CONTENTS

## ECOLOGY: EARTH'S LIVING RESOURCES

## *Activity Bank/Reference Section*

## *Features*

# CONCEPT MAPPING

Throughout your study of science, you will learn a variety of terms, facts, figures, and concepts. Each new topic you encounter will provide its own collection of words and ideas—which, at times, you may think seem endless. But each of the ideas within a particular topic is related in some way to the others. No concept in science is isolated. Thus it will help you to understand the topic if you see the whole picture; that is, the interconnectedness of all the individual terms and ideas. This is a much more effective and satisfying way of learning than memorizing separate facts.

Actually, this should be a rather familiar process for you. Although you may not think about it in this way, you analyze many of the elements in your daily life by looking for relationships or connections. For example, when you look at a collection of flowers, you may divide them into groups: roses, carnations, and daisies. You may then associate colors with these flowers: red, pink, and white. The general topic is flowers. The subtopic is types of flowers. And the colors are specific terms that describe flowers. A topic makes more sense and is more easily understood if you understand how it is broken down into individual ideas and how these ideas are related to one another and to the entire topic.

It is often helpful to organize information visually so that you can see how it all fits together. One technique for describing related ideas is called a **concept map**. In a concept map, an idea is represented by a word or phrase enclosed in a box. There are several ideas in any concept map. A connection between two ideas is made with a line. A word or two that describes the connection is written on or near the line. The general topic is located at the top of the map. That topic is then broken down into subtopics, or more specific ideas, by branching lines. The most specific topics are located at the bottom of the map.

To construct a concept map, first identify the important ideas or key terms in the chapter or section. Do not try to include too much information. Use your judgment as to what is really important. Write the general topic at the top of your map. Let's use an example to help illustrate this process. Suppose you decide that the key terms in a section you are reading are School, Living Things, Language Arts, Subtraction, Grammar, Mathematics, Experiments, Papers, Science, Addition, Novels. The general topic is School. Write and enclose this word in a box at the top of your map.

SCHOOL

Now choose the subtopics—Language Arts, Science, Mathematics. Figure out how they are related to the topic. Add these words to your map. Continue this procedure until you have included all the important ideas and terms. Then use lines to make the appropriate connections between ideas and terms. Don't forget to write a word or two on or near the connecting line to describe the nature of the connection.

Do not be concerned if you have to redraw your map (perhaps several times!) before you show all the important connections clearly. If, for example, you write papers for Science as well as for Language Arts, you may want to place these two subjects next to each other so that the lines do not overlap.

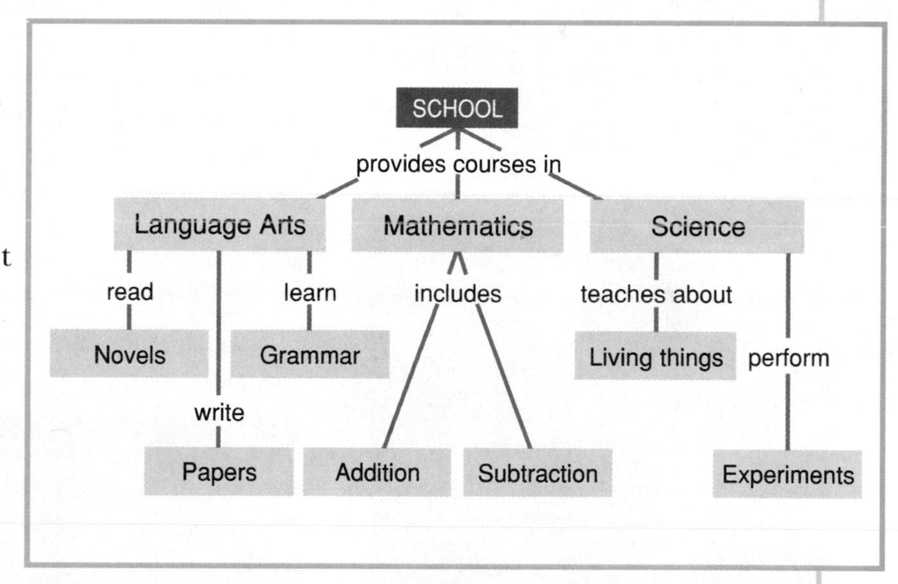


One more thing you should know about concept mapping: Concepts can be correctly mapped in many different ways. In fact, it is unlikely that any two people will draw identical concept maps for a complex topic. Thus there is no one correct concept map for any topic! Even though your concept map may not match those of your classmates, it will be correct as long as it shows the most important concepts and the clear relationships among them. Your concept map will also be correct if it has meaning to you and if it helps you understand the material you are reading. A concept map should be so clear that if some of the terms are erased, the missing terms could easily be filled in by following the logic of the concept map.

# ECOLOGY

## Earth's Living Resources

*The red-cockaded woodpecker has an amazing way of protecting its young from the red rat snake.*

The rat snake's tongue flickers in and out of its mouth, tasting the air. Food is nearby—warm, fat baby birds! The snake slithers up an old, diseased pine tree in search of its prey. But before it can reach the cavity in the tree trunk where the nest is hidden, the snake slithers into an unexpected patch of sticky pine sap. The sap gums up the snake's scales so that it loses its grip and falls off the tree.

A few minutes later, a red-cockaded woodpecker flies into the cavity in the old pine tree. After feeding its babies, the woodpecker does some housekeeping. Clinging to the trunk of the tree, the woodpecker pecks several holes

*Many different kinds of places are home to Earth's living things. This forest of pink rhododendrons, Douglas firs, and redwoods is located in California's Redwoods State Park.*

# CHAPTERS



into the bark surrounding the cavity. Sap runs out of the holes and oozes slowly down the tree. Then the woodpecker flies off to gather more food for its hungry babies.

The pine tree, snake, and woodpeckers you have just read about interact in an interesting way. In this textbook, you will first learn about the different kinds of interactions that occur among living things and between living things and their nonliving surroundings. Next, you will read about life cycles and other patterns of change in nature. You will then learn about the basic kinds of places that are home to Earth's living things. Finally, you will explore the reasons why organisms such as the red-cockaded woodpecker are in danger of disappearing forever from the Earth.

*Living things interact in many different ways. Locking horns to establish who's the boss, these two bull moose are battling in Alaska's Denali National Park.*

## Discovery Activity

### Seeds of Change

1. Collect as many different kinds of uncooked seeds as you can from the foods you eat.
   - What kinds of foods contain or are made up of seeds?
   - Where are these foods grown?
2. Obtain a paper towel and a small glass jar with a lid. Fold the paper towel so that it can line the sides of the jar. Moisten the paper towel with water. Then place some of the seeds in the jar so that they are sandwiched between the glass and the paper towel. Cover the jar.
3. Observe the seeds daily.
   - How do the seeds change over time?
   - How might new kinds of plants appear in an area?
   - What sorts of things might affect what happens to the seeds?

# Interactions Among Living Things

## Guide for Reading

*After you read the following sections, you will be able to*

**1–1 Living Things and Their Environment**

- Define ecosystem and identify the parts of an ecosystem.

**1–2 Food and Energy in the Environment**

- Discuss the interactions among producers, consumers, and decomposers.
- Describe how energy flows within an ecosystem.

**1–3 Interaction and Evolution**

- Discuss how interactions affect evolution.

**1–4 Life in the Balance**

- Explain why a disturbance in the balance in one part of an ecosystem can affect the entire ecosystem.

As sunlight falls on the leaves of a plant, substances in the leaves capture the sunlight's energy and use it to make food. But even as the leaves are making food, a tiny thief is stealing some of it. A small aphid (an insect) pokes its strawlike mouthparts into the leaf and begins to suck up food-rich sap.

Suddenly, a hungry ant scurries along the leaf toward the aphid. Is the aphid doomed to end up as the ant's lunch? No. Upon reaching the aphid, the ant begins to stroke the smaller insect with its feelers. The aphid responds by releasing a drop of a sugary substance called honeydew. The ant eagerly licks up the honeydew. Then the ant gently picks up the aphid in its jaws and carries it to another leaf. There the aphid is added to a "herd" being tended by ants. The ants take care of the aphids in exchange for meals of honeydew. The ants move the aphids to fresh leaves when the old ones wither. When it rains, the ants carry the aphids to more sheltered leaves. The ants also defend their herd from ladybugs and other aphid-devouring animals.

The interactions among aphids, ants, ladybugs, sunlight, and plants are just a few of the countless relationships that link living things to one another and to their surroundings. Read on to discover more about interactions among living things.

### Journal *Activity*

***You and Your World*** In your journal, explore the thoughts and feelings you have about environmental issues.

*These orange-colored ants are busily tending a large "herd" of dark gray aphids.*

**Guide for Reading**

*Focus on these questions as you read.*

- *Why do ecologists study both the living and the nonliving parts of an environment?*
- *What do the following terms mean: environment, ecosystem, community, population, and habitat?*

# 1–1 Living Things and Their Environment

A thousand meters below the ocean's surface, the last traces of sunlight fade into nothingness. The water temperature is only a few degrees above freezing. There is very little food or oxygen. Yet the harsh, dark world of the deep sea is home to many organisms (living things). Nightmarish fishes with huge teeth and eyes glow with ghostly lights made by their own bodies. Octopuses and squids with webbed arms pulse through the water or float like falling parachutes. Strange spiky sea cucumbers sift through the muddy ocean floor for the bits of food that drift down from the sunlit world above.

On land, in a lush tropical rain forest, tall trees with clinging vines thrive in the warmth and sunlight. In the treetops, brightly colored parrots munch on seeds while monkeys chatter to one another. Snakes and lizards climb up and down the tree trunks in search of food. And piglike tapirs calmly make their way along the ground.

The deep sea and a rain forest are only two of the many different **environments** found on Earth. An

**Figure 1–1** *The huge jaws and enormous teeth of this deep-sea fish enable it to catch and eat animals that are larger than itself. The eyelash viper and the* Heliconia *flowers on which it is coiled live in the same kind of environment as the blue-and-gold and scarlet macaws. What kind of environment is home to these organisms?*

environment consists of all the living and nonliving things with which an organism may interact.

Organisms obtain the food, water, and other resources they need to live and grow from their environment. Consider for a moment some of the things a parrot gets from its rain-forest environment. The parrot feeds on seeds and fruits from plants. It drinks water from puddles and streams. It has trees in which to perch and build its nest. It has air to breathe and to fly through. The parrot can live in a rain forest because this environment contains all the things a parrot needs to survive. Why wouldn't a parrot be able to live in an environment that is quite different from the rain forest, such as the deep sea or a desert? Can you explain why different environments contain different kinds of organisms?

**Figure 1–2** *A few moments ago, the chameleon's environment was changed by the appearance of a juicy cricket. How does the chameleon respond to such short-term changes in its environment?*

Living things do not simply exist in their environment like photos in a frame. They constantly interact with their environment. Organisms can change in response to conditions in the environment. These changes can often be quite rapid. For example, within a few seconds, the fish known as a flounder can change its colors and spots to match the sand and pebbles on a new patch of ocean floor. Or the changes can be much slower. On windswept mountains, for example, trees grow so that they bend in the direction of the wind. Some slow changes involve entire groups of organisms, not just individual organisms. In Section 1–3, you will read about some of the ways organisms have evolved (changed over time) in response to their environment.

In addition to changing in response to their environment, living things also cause change in their environment. Earthworms and other burrowing animals dig tunnels in the soil. Woodpeckers drill holes in trees. Tree roots break up sidewalks. Beavers build dams that block flowing streams and thus create ponds. Can you identify some ways in which humans change their environment?

**Figure 1–3** *The relationships among the living and nonliving parts of the environment can be thought of as being like a giant spider web. But an environmental web is more complex—and perhaps more fragile—than the delicate web of a spider.*

**All of the living and nonliving things in an environment are interconnected.** You can think of an environment as being like a giant spider web. However, the threads of this web are not spun from silk. The threads of an environment's web are the relationships among its plants, animals, soil, water, temperature, light, and other living and nonliving things.

Think for a moment about what happens when an insect gets caught in a spider's web. As one thread of the web is disturbed, the shaking motion is transferred to all the threads that are part of the web. In an environmental web, changes in one thread may also be transmitted to other threads and have an effect on them. For example, cutting down the trees in a forest may affect the rainfall in a distant city. And when a thread is broken, the entire web is weakened.

To understand the changes that can occur in an environment and how they can affect the environment, you can study the science called **ecology.** Ecology is the study of the relationships and interactions of living things with one another and with their environment. Scientists who study these interactions are called ecologists.

**Figure 1–4** *A forest of broad-leaved trees is just one of the many kinds of ecosystems on Earth. What kinds of organisms would you find in this ecosystem?*

## Ecosystems

Living things inhabit many environments on Earth. From the polar ice caps to the forests and plains of the equator, living things can be found under ground, in air, in water, and on land. Organisms have been found at the bottom of ocean trenches kilometers deep and floating in the air more than eight kilometers above the Earth's surface.

To make sense of the number and variety of interactions among Earth's living things and their environment, ecologists find it useful to divide the world up into separate units known as **ecosystems.** An ecosystem consists of all the living and nonliving things in a given area that interact with one another. A forest ecosystem, for example, includes birds and squirrels in the trees, foxes and rabbits in the bushes, the trees and bushes themselves, insects and spiders, shade-loving wildflowers, ferns, mushrooms and other fungi, microorganisms (microscopic organisms) such as bacteria and protists, dead leaves, chemicals in the soil, rocks, sunlight, rain water, and many other living and nonliving things. How do the trees in a forest ecosystem interact with squirrels? With the soil?

An ecosystem can be as tiny as a drop of pond water or a square meter of a garden. Or it can be as large as an ocean, a forest, or a planet. The size of

an ecosystem is defined by the ecologist who is studying it. What ecosystem would you expect an ecologist interested in the interactions of freshwater swimming microorganisms to study? Would an ecologist studying the wildlife in and around Lake Tahoe, California, use a similar ecosystem? Why or why not?

It is useful to talk about ecosystems as if they were separate, self-contained units. However, it is important to keep in mind that ecosystems are not isolated. Ecosystems overlap and affect one another. The grizzly bears of a forest ecosystem may feed on the salmon of a stream ecosystem. Chemicals from aerosol cans, air conditioners, and refrigerators in the United States and elsewhere are carried great distances by the wind and eventually break down the protective ozone layer in the air above the poles. The "holes" thus created in the ozone layer allow extra radiation from the sun to reach the ecosystems near the poles, damaging them. Damage to these ecosystems may, in turn, result in damage to other ecosystems—including ones in which you live! So it is important that you realize all the living and nonliving things on Earth are ultimately connected to one another.

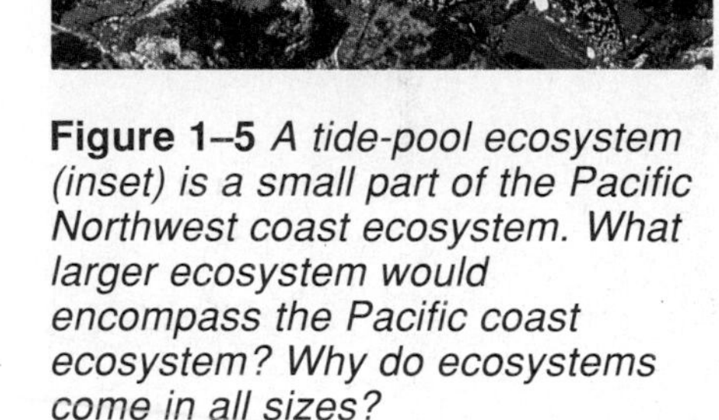

**Figure 1–5** *A tide-pool ecosystem (inset) is a small part of the Pacific Northwest coast ecosystem. What larger ecosystem would encompass the Pacific coast ecosystem? Why do ecosystems come in all sizes?*

*Identifying Interactions*

Ecosystems are all around you. Choose one particular ecosystem—an aquarium, swamp, lake, park, or city block, for example—and study the interactions that occur among the living and nonliving parts of the ecosystem. Include drawings and diagrams in your observations.

## Communities

The living part of any ecosystem—all the different organisms that live together in that area—is called a **community.** The community of a pond, for example, might include fishes, frogs, snails, microorganisms, and water lilies. The members of a community interact with one another in many different ways. Lily pads provide a resting place for frogs. Large fishes

Figure 1–6 *The plains of Africa are home to many different kinds of living things. What members of the African plains community can you identify here?*

Figure 1–7 *Some populations—such as that of the flamingoes in Botswana, Africa—are enormous. Others—such as that of the treehoppers on a twig in Costa Rica—are quite small.*

eat frogs. Microorganisms break down the bodies of dead organisms, producing products such as nitrogen compounds that can be used by plants. Can you think of some other ways in which the members of a pond community interact? In the next two sections, you will learn about some of the specific interactions that take place within communities.

## Populations

You, like all other living things on Earth, belong to an ecological community. Your community probably contains many different kinds of living things: people, dogs, cats, birds, insects, grass, and trees, to name a few. What other kinds of organisms are found in your community?

Each kind of living thing makes up a **population** in the community. A population is a group of organisms of the same type, or species, living together in the same area. (A species is a group of similar organisms that can produce offspring. You and all other humans belong to the same species. But cats belong to another species.) For example, all the rainbow trout living in a lake are a population. All the redwood trees in a forest are a population.

But a group consisting of all the wildflowers in a meadow is not considered to be a population. Can you explain why?

## Habitats

Where would you go to find a lion? How about a pigeon? Where in a forest would you look for a squirrel? A mushroom? An earthworm? Would you discover all these organisms in the same place? Probably not. Lions live on the grassy plains of Africa. Pigeons live in cities, among other places. In a forest, squirrels live in the trees, mushrooms grow on the forest floor, and earthworms burrow in the soil. Each of these organisms lives in a different place.

The place in which an organism lives is called its **habitat.** A habitat provides food, shelter, and the other resources an organism needs to survive. Living things such as lions, pigeons, and mushrooms live in different habitats because they have different requirements for survival. Organisms such as lions, zebras, and giraffes also have different requirements for survival. Yet these organisms live in the same habitat. Why? Because their requirements—such as for temperature, water, and open space—overlap in many ways. The size of an organism's habitat depends on the organism's habits and needs. The habitat of a humpback whale is the open ocean. The habitat of a certain tiny mite, on the other hand, is the ear of a moth.

### ACTIVITY DISCOVERING

*Home Sweet Home*

**1.** Choose one of the following animals and find out what kind of shelter it builds: beaver, trapdoor spider, mud dauber wasp, prairie dog, cliff swallow, termite, weaver bird, mole rat, toucan, carrier shell (*Xenophora*), coral gall crab.

**2.** On a sheet of paper, draw a picture of the animal's shelter and make a list of the materials needed to build it.

**3.** Build a model of your animal's shelter using the same materials the animal would use whenever possible.

■ Predict what would happen if the materials that an animal uses to build its shelter were not available.

## 1–1 Section Review

1. Why do ecologists study both the nonliving and living things in an environment?
2. What is an ecosystem? Give an example of an ecosystem.
3. What is the difference between a community and a population?

**Connection—*Architecture***

4. Explain why an architect designing a new home for the cheetahs in a zoo must know something about the natural habitat of cheetahs.

**Guide for Reading**

*Focus on these questions as you read.*

- *How do producers, consumers, and decomposers interact?*
- *What is the difference between a food chain and a food web?*

# 1–2 Food and Energy in the Environment

If you enjoy watching or playing team sports, you know that the members of a team usually play different positions. Basketball players may be centers, forwards, or guards. Baseball players may be pitchers, catchers, shortstops, center fielders, and so on. Each position has a particular role associated with it. For example, a pitcher throws the ball to the batters. A guard tries to prevent the members of the other team from scoring a basket. Similarly, organisms have special roles that they play in an ecosystem.

## Energy Roles

Organisms may be **producers, consumers,** or **decomposers.** These three terms indicate how an organism obtains energy and how it interacts with the other living things in its community.

**PRODUCERS** Some organisms, such as green plants and certain microorganisms, have a very special ability that sets them apart from all other living things: They can make their own food. Such organisms are known as producers. Producers are able to use a source of energy (such as sunlight) to turn simple raw materials (such as water and carbon dioxide gas) into food (such as the sugar glucose). Organisms that cannot make their own food may eat the producers directly. Or they may eat other organisms that cannot make their own food. However, all organisms that cannot make their own food ultimately depend on producers. **Producers are the source of all the food in an ecosystem.**

**Figure 1–8** *Towering redwood trees, indigo Texas bluebonnets, and bright pink phlox are examples of producers. Why are producers essential for life on Earth?*

Figure 1–9 *The caterpillars, shark, and hyena and vulture are all consumers.*

**CONSUMERS** Organisms that cannot make their own food depend on producers for food and energy. **An organism that feeds directly or indirectly on producers is called a consumer.**

There are many kinds of consumers. Some organisms, such as grasshoppers and rabbits, are plant eaters. Plant eaters are known as herbivores. The term herbivore comes from the Latin words *herba,* which means grass or herb, and *vorare,* which means to eat. Spiders, snakes, and wolves, which eat other animals, are known as carnivores. The Latin word *carnis* means of the flesh. Why is the term carnivore appropriate for organisms that eat meat?

Organisms that eat both plants and animals are known as omnivores. (The Latin word *omnis* means all.) Crows, bears, and humans are just a few examples of omnivores.

There are many other terms that are used to group consumers according to what they eat. One such term is scavenger. A scavenger is an animal that feeds on the bodies of dead animals. Jackals, hyenas, and vultures are examples of scavengers. So are certain crayfish and crabs, who "clean up" watery environments by eating dead organisms.

**DECOMPOSERS** After living things die, organisms called decomposers use the dead matter as food. **Decomposers break down dead organisms into simpler substances.** In the process, they return important

*Diet Delights*

**1.** Use a dictionary to find out what these specialized feeders eat: piscivore, insectivore, detritovore, myrmecophage, frugivore, coprophage, necrophage, nectarivore, granivore, apivore, carpophage.

**2.** Invent a term to describe the feeding habits of Count Dracula (and other vampires).

Activity Bank

Garbage in the Garden, p.138

**Figure 1–10** *Nestled among the fallen leaves on the forest floor, these mushrooms are slowly breaking down a dead branch for food. Why are mushrooms considered to be decomposers?*

materials to the soil and water. You may be familiar with the term "decay," which is often used to describe this process. Molds, mushrooms, and many kinds of bacteria are examples of decomposers.

Decomposers are essential to the ecosystem because they rid the environment of the bodies of dead plants and animals. Even more importantly, decomposers return nutrients (compounds containing chemicals such as nitrogen, carbon, phosphorus, sulfur, and magnesium) to the environment. These nutrients are then used by plants to make food, and the cycle of nutrients through the environment continues. If the nutrients were not returned to the environment, organisms within that ecosystem could not survive for long.

## Food Chains and Food Webs

In general, food and energy in an ecosystem flow from the producers to the consumers, and finally to the decomposers. The food and energy links among the producers and consumers in an ecosystem are represented by **food chains** and **food webs.**

A food chain represents a series of events in which food and energy are transferred from one organism in an ecosystem to another. The first link in a food chain is always a producer. The second link is a herbivore. The third link and all the links after that are almost always carnivores.

Let's take a look at an example of a food chain. In an Antarctic food chain, the producers are one-celled organisms known as diatoms. The diatoms capture energy from the sun and use it to make food. When a diatom is eaten by a shrimplike animal called a krill, the food energy and matter in the diatom are transferred to the krill. In the following links of the food chain, the krill is eaten by a squid, which is eaten by a penguin, which is eaten by a leopard seal, which is eaten by a killer whale. Both food energy and matter are transferred at each successive link of the food chain. Figure 1–11 illustrates another food chain.

The "end" of a food chain is connected to the "beginning" by decomposers. In the Antarctic food chain, decomposers break down the body of the killer whale when it dies. This makes matter in the

### ACTIVITY DOING

*Your Place in Line*

Every time you eat, you are assuming a particular place in a food chain. You can determine your place in a food chain through the following activity.

Make a list of all the foods you ate for breakfast, lunch, and dinner. For each food, figure out the probable links in the food chain that lead up to you. Be sure to answer the following questions:

1. Who are the producers?
2. Who are the consumers?

form of nutrients available to the producers. What do the producers do with these nutrients?

A food chain gives you a glimpse of the food and energy relationships in an ecosystem. But it does not give you the whole picture. There are many organisms in an ecosystem, and few of them eat only one kind of food. Thus there must be more than one food chain in an ecosystem. Figure 1–12 on page 22 shows how a number of organisms in the Antarctic ecosystem are linked by food and energy relationships. This kind of diagram is known as a food web. Can you see why the name food web is an appropriate one? A food web consists of many overlapping food chains. One of the food chains in this food web was just described. Another food chain might be: a diatom is eaten by a tiny water animal that is eaten by a fish that is eaten by a penguin. Take a moment now to identify three of the many other food chains in this food web.

**Figure 1–11** *The links of this desert food chain include a flowering Saguaro cactus (left), an iguana (top right), and a roadrunner (bottom right). What role does each organism play in this food chain?*

## Feeding Levels and Energy

A feeding level is the location of an organism along a food chain. Producers form the first feeding level. Herbivores form the second feeding level. And carnivores form the third feeding level.

At each feeding level, organisms use the energy they obtain to digest their food, reproduce, move,

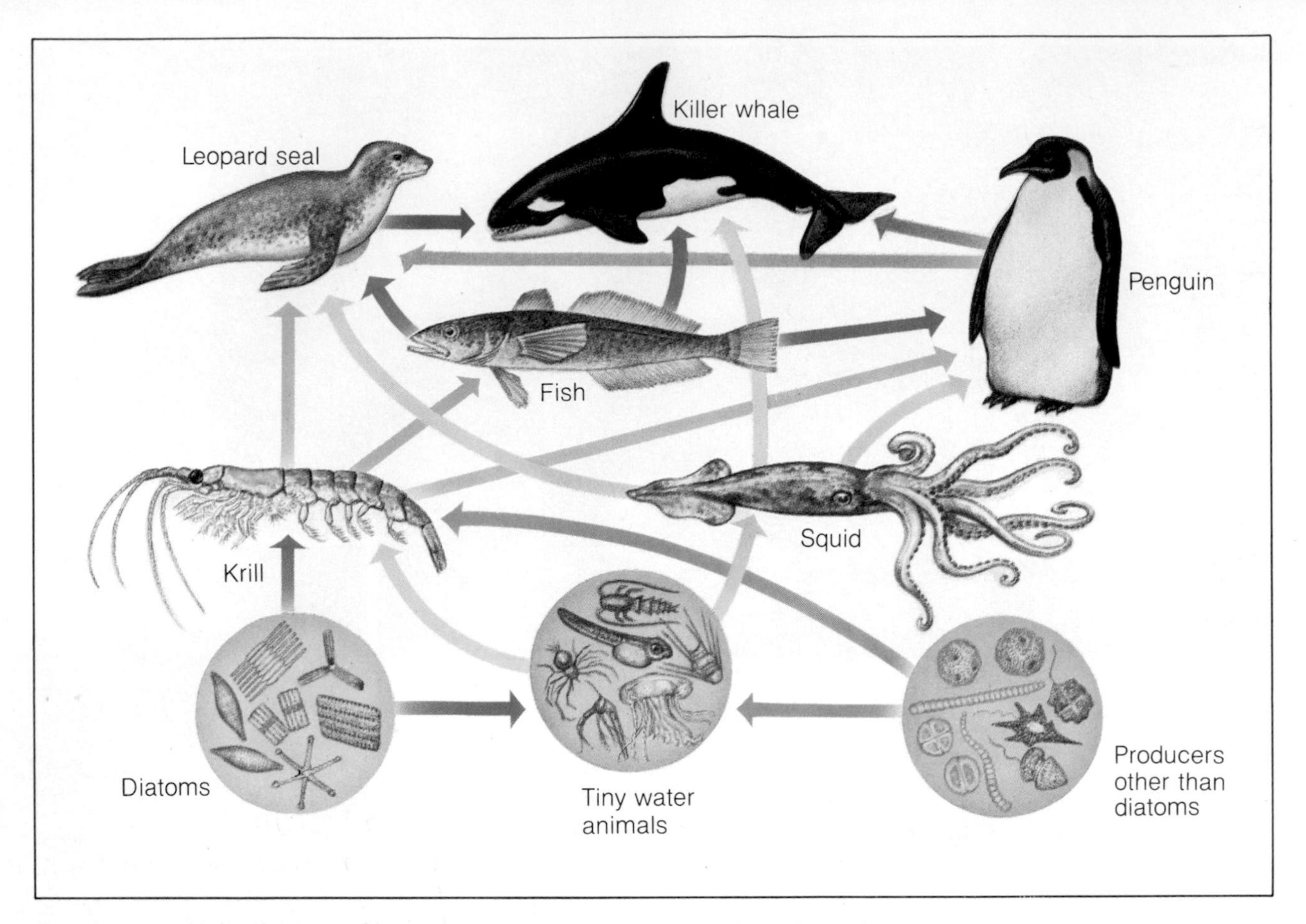


**Figure 1–12** *Food webs can be quite complicated, even when they show only a few of the organisms in an ecosystem. What are the producers in this Antarctic food web? What are the herbivores and carnivores?*

grow, and carry out other life activities. What does this mean for living things at higher feeding levels? It means that there is less energy available to them. As you can see in Figure 1–13, the amount of energy at the second feeding level is much smaller than that at the first feeding level. The amount of energy at the third feeding level is smaller still. At each successive level, there is less energy than there was before. Because a diagram that shows the amount of energy at the different feeding levels looks like a pyramid—wide at the base and narrowing toward the top—it is called a pyramid of energy.

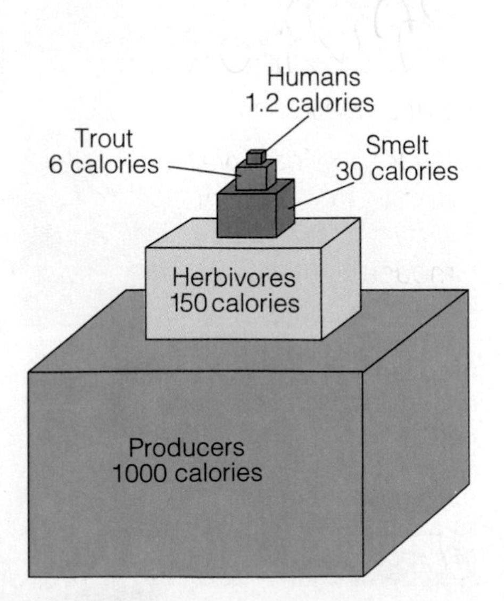


**Figure 1–13** *With each successive feeding level, the amount of energy decreases greatly. How does the amount of energy in trout compare to that in smelt? Why can more people be fed if smelt is eaten instead of trout? Why do some experts think that a first step in solving the problem of world hunger might be to encourage people to eat lower on the food chain?*

## 1–2 Section Review

1. Explain how producers, consumers, and decomposers interact. For each energy role, give two examples of organisms that perform that role in their ecosystem.
2. What is a food chain? A food web?
3. What is the relationship between feeding levels and amount of available energy?
4. How does energy flow through an ecosystem?

**Critical Thinking—*Evaluating Diagrams***

5. Why are decomposers left out of almost all diagrams of food chains and food webs? Do you think this is a good practice? Why or why not?

# 1–3 Interaction and Evolution

**Guide for Reading**

*Focus on this question as you read.*

▶ *What are competition, predation, and symbiosis?*

Each organism in a community has its own unique role to play. This role, or **niche** (NIHCH), consists of more than the organism's place in a food chain. It includes everything the organism does and everything the organism needs in its environment. In other words, an organism's niche includes the place in which it lives, the food it eats, the organisms that feed on it, the organisms that interact with it in other ways, the amount of light and humidity it needs, and the physical conditions in which it can survive.

As you learned in the previous section, two or more species (kinds of organisms) can share the same habitat, or place in which to live. For example, the coral reef habitat is home to corals that build stony skeletons, soft flowerlike sea anemones, flat ribbons and thin threads of seaweed, spiny sea urchins, scuttling crabs, fishes in a rainbow of colors, and many other organisms. Two species can also share similar habits and food requirements. For example, flying fox bats and toucans are both flying animals that live in trees and eat fruit. Sharks and

*Eat or Be Eaten*

Draw a food web that contains the following: bread crumbs, food scraps, pigeon, mouse, cockroach, cat, rat, bacteria, starling, spider, fly. Where would you be likely to find this food web? What feeding level is missing from this food web? Why is this feeding level missing?

Figure 1–14 *The flying fox bat of Australia and the Toco toucan of Brazil both enjoy feasting on fruit. How are their niches similar? How are they different?*

dolphins are fast-swimming ocean animals that both eat fish. Two or more species *cannot* share the same niche. If two species try to occupy the same niche for a long period of time, one of the species will become extinct, or die off. Why? The answer has to do with a special kind of relationship between organisms—a relationship you are now going to read about.

## Competition

Ecosystems cannot satisfy the needs of all the living things in a particular habitat. There is only a limited amount of food, water, shelter, light, and other resources in an environment. Because there are not enough resources to go around, organisms must struggle with one another to get the things they need to survive. This type of interaction is known as **competition.**

One of the resources for which organisms compete is food. The shortage of other resources, such as water, light, and suitable places to live, also results in competition. Regardless of the specific causes, competition can have a powerful effect on the size and location of a population in an ecosystem.

Competition can occur within a species as well as between species. A male lion will fight with another male lion for control over a group of female lions. Pairs of penguins squabble with other pairs of penguins over the pebbles used in nest building. And if there is not enough space between two plants of the same species, they will compete with one another for

*Shady Survival*

Suppose 140 trees are growing in a shaded area of a forest. Of those trees, 65 percent are sugar maples and 35 percent are red maples. How many trees of each kind are there?

**Figure 1–15** *Bighorn rams knock their horns together with a resounding crash as they compete to determine which one will gain control of a group of females.*

water, nutrients, and light. (This is why gardeners have to thin out seedlings in a flower bed or vegetable patch.)

In order to better understand how competition works in nature, it might be helpful for you to think about some more familiar kinds of competition. People who read for a play, audition for a musical group, or try out for a sports team compete with one another for limited resources: roles, chairs, and positions, respectively. Nonliving things can also be thought of as being in competition with one another. Have you ever walked down the aisle of a supermarket and gazed in amazement at the number of different kinds of breakfast cereal displayed? Like living things, products such as breakfast cereals, cars, shampoos, blue jeans, and even science textbooks compete with one another for a limited resource. In this case, the resource is a customer's purchase choice. The products that are most successful at obtaining this resource (the ones that are bought most often) continue to be sold. The products that are least successful soon cease to be manufactured.

## Predation

Slowly and silently, the cat sneaks up on an unsuspecting mouse. The cat crouches down. The tip of its tail twitches and its muscles tense. Then the cat pounces!

For thousands of years, people have valued the cat's ability to kill rats and mice. Living things, such as cats, that catch, kill, and eat other living things

**Activity Bank**

On Your Mark, Get Set, Grow!, p.141

*"Avenues of Delight and Discovery"*

The works of Rachel Carson (1907–1964) prove that scientific writings can be as beautiful and poetic as any work of literature. Although she was a marine biologist by training, two of her most famous books are not about the sea and its creatures. Read *The Silent Spring* and *The Sense of Wonder*, by Rachel Carson. Explore the beauty and mystery of the natural world by simply making a trip to the library!

**Figure 1–16** *How can you tell that the lioness (right), the great blue heron (top left), and the hawk (bottom left) are predators? What kinds of prey do these predators hunt? By the way, the attacking redwing blackbird is not trying to prey on the hawk. The hawk simply flew too close to the blackbird's nest, upsetting the smaller bird.*

are called **predators.** The organisms that are eaten by predators are called **prey.** Although people usually think of predators and their prey as being animals, ecologists consider just about all situations in which one organism kills and eats another as examples of predation.

Predation, like competition, plays an important part in shaping the structure of communities. In catching and eating prey, predators help to reduce the size of prey populations. By helping to control the size of prey populations, predators also help to maintain the diversity in an ecosystem. When predators are absent, prey species can become too numerous and crowd out other organisms.

This is exactly what happened with the rabbits that were introduced to Australia about 150 years ago by European settlers. The rabbits had few predators, so their population increased very quickly. These harmless-looking animals were soon stripping the grasslands bare of vegetation. Many of the native animals that also fed on this vegetation, such as kangaroos, starved. To save their herds, cattle ranchers put up special fences to keep the rabbits out of their pastures. This situation continued until the 1950s, when scientists introduced a rabbit-killing predator to Australia: a virus that caused a fatal rabbit disease. The disease killed off about 80 percent of the rabbits, making it possible to reclaim land for native Australian wildlife and livestock.

# PROBLEM Solving

## *To Bee or Not to Bee*

The honeybees in the United States are in trouble! Pests from distant lands have been accidentally introduced into this country. These pests now threaten the $150-million-a-year beekeeping industry. Agricultural researchers, bee specialists, and other experts are searching frantically for a way to fight these pests before it's too late. Can you help them find a solution to the problem of the pests?

For each of the following situations, (1) identify the type of interaction involved, (2) describe the problem you perceive and propose a solution, and (3) predict the effects that the situation may have on the further evolution of the honeybee.

**Evaluating situations**

**1.** Tracheal mites live inside a bee's breathing tubes. There they suck fluids from the bee's body and interfere with its breathing. American honeybees are descendants of European honeybees, which rarely have tracheal mites. However, tracheal mites have been a major problem for American honeybees since 1984.

**2.** *Varroa* mites, which were first discovered in the United States in September 1987, attach to the outside of adult bees and young bees. The mites weaken the bees by feeding on their body fluids. In addition, they may transmit diseases from one bee to another.

**3.** Although "killer bees" belong to the same species as American honeybees, they are much more aggressive. Killer bees were accidentally released in Brazil in 1957. Since then, they have been expanding their range northward. In October 1990, the first swarm of killer bees entered the United States.

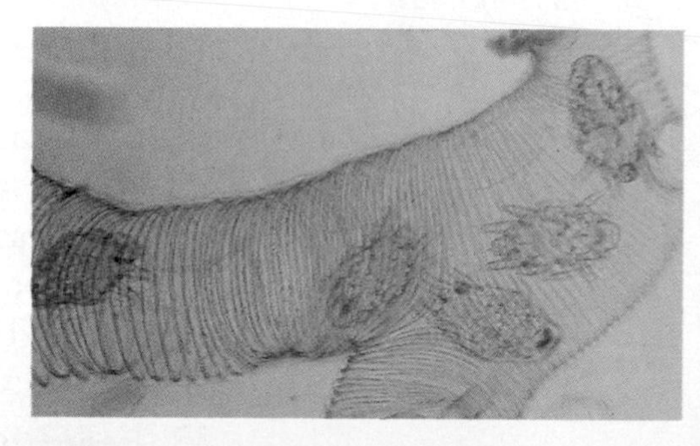

## Symbiosis

**Figure 1–17** *The ratel loves honey, but cannot easily find beehives (top). The honeyguide can easily find beehives, but is too small, weak, and vulnerable to bee stings to get at the beeswax it likes to eat (bottom). Together, the ratel and honeyguide can obtain their favorite treats. What is this kind of partnership called?*

In a tropical ocean, a remora fish uses a structure on top of its head to attach itself to the belly of a shark and get a free ride. Chirping and fluttering, a honeyguide bird in Africa leads a furry black-and-white ratel to a wild beehive. With its sharp claws, the ratel rips open the hive. The ratel then laps up the honey as the honeyguide bird dines on beeswax. In the United States, a dog sleeping on a rug suddenly sits up and begins to scratch an itchy flea bite.

What do these events have in common? They are all examples of **symbiosis** (sihm-bigh-OH-sihs; plural: symbioses). **Symbiosis is a close relationship between two organisms in which one organism lives near, on, or even inside another organism and in which at least one organism benefits.** Symbioses are placed into three categories: **commensalism, mutualism,** and **parasitism.** In commensalism, one of the organisms benefits and the other is not harmed by the association. In mutualism, both organisms benefit. And in parasitism, one organism benefits and the other is harmed. Is the relationship between a dog and a flea an example of commensalism, mutualism, or parasitism? What kind of symbiosis do you have with a pet animal?

As you read the following examples of commensalism, mutualism, and parasitism, keep in mind that science is a constantly changing body of knowledge. New observations may lead scientists to conclude that what was thought to be one kind of symbiosis is in fact another kind altogether. Try to think of additional ways in which the partners in a symbiosis may help—or harm—each other.

**COMMENSALISM** High in the branches of a tree, a large, fierce hawk called an osprey builds a big, flat nest for its eggs. Smaller birds, such as sparrows and wrens, set up their homes beneath the osprey's nest. Because the osprey eats mostly fish, these smaller birds are in no danger from the osprey. In fact, the little birds obtain protection from their enemies by living close to the fierce hawk.

Beautiful orchids and exotic bromeliads (relatives of pineapples) survive in dense, shadowy jungles by growing on tall trees. There among the tree branches, the plants get a great deal of sunlight. The roots

of these plants are exposed, so they can take water and nutrients right out of the air or off the surface of the tree's bark. Can you explain why the relationship between a bromeliad and a tree is an example of commensalism?

Like a giant tropical tree, you have smaller members of your community living on you. Dozens of tiny mites live at the base of the hairs that make up your eyebrows. Don't rush to the mirror to check—these mites can be seen only with a microscope. And don't worry about being a home to these mites. They are quite harmless, and everybody has them!

**Figure 1–18** *Sea anemones use stinging tentacles to catch and stun small fishes and other prey. This clownfish, however, is immune to the anemone's sting and thus is safe in the anemone's deadly embrace (left). The bromeliads that grow on forest trees are in turn involved in commensalism with smaller creatures, such as this red-eyed tree frog, that make their home in the vaselike center of the plant (right).*

**MUTUALISM** Remember the ants and aphids you read about at the beginning of this chapter? This relationship is one of the many examples of mutualism that exist among Earth's living things.

The goby fish and snapping shrimp shown in Figure 1–19 on page 30 live in a sandy burrow built by the shrimp. Because the shrimp is nearly blind, it keeps one of its long feelers on the goby. When danger approaches, the goby warns its partner with flicks of its fins, and both partners retreat to their burrow. Neither the goby nor the shrimp can survive on its own. Gobies without homes and shrimp without guides are soon eaten up by predators.

Some of the most important (if not the most interesting) examples of mutualism involve microorganisms that live inside the body of much larger partners. The microorganisms obtain a safe home

**Figure 1–19** *In mutualism, both organisms benefit from the symbiosis. The small brown oxpeckers on the rhino's nose and belly feed on ticks and other parasites (right). (The white cattle egrets are involved in commensalism with the rhino—they eat insects flushed from hiding as the rhino walks through the grass.) The honeybee is extending a tubelike mouthpart to sip nectar (a sugary liquid) from a flower (top left). The yellow dust on the bee's head is pollen. Most flowers need to receive pollen from another flower in order for their seeds to develop. The goby and snapping shrimp depend on each other for survival (bottom left).*

inside their partner. In return, they help their partner in some way. Bacteria that live inside your digestive system help to produce certain vitamins that your body needs. Microorganisms in certain species of water animals (deep-sea tube worms, giant clams, corals, flatworms, and sea slugs, to name a few) produce food for their partners. Microorganisms in the intestines of cattle, horses, rabbits, termites, and other herbivores help these animals to digest the tough plant materials that they eat. Flashlight fish have structures beneath their eyes that contain symbiotic bacteria. These bacteria produce light, which the flashlight fish are able to turn on and off. The blinking lights of a school of flashlight fish are thought to confuse predators and perhaps enable the fish in the school to communicate.

**PARASITISM** Have you ever had a cold or been bitten by a mosquito or flea? If so, you have had firsthand experience with parasitism.

Parasites come in many shapes and sizes. Blood-drinking animals such as fleas, ticks, mosquitoes, leeches, and vampire bats are parasites. The fungi that cause athlete's foot and ringworm are also parasites. Some parasites live inside the body of another organism. Disease-causing bacteria and viruses are internal parasites. So are a number of worms, including the heartworm that affects pet dogs. Although they vary greatly in habit and appearance, all parasites have one trait in common: They are involved in a symbiotic relationship in which they harm their partner.

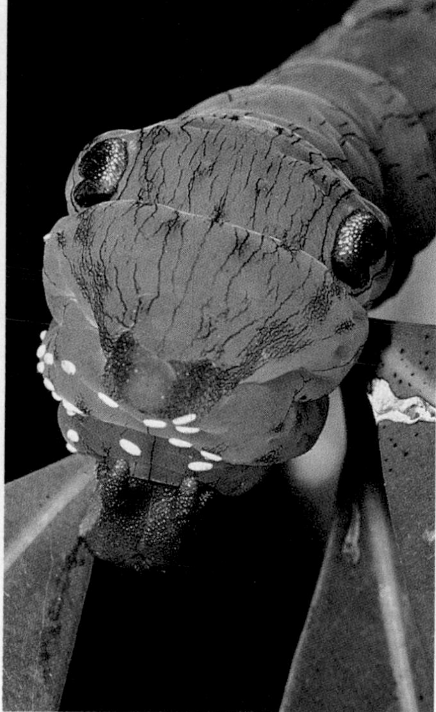

Ecologists often regard parasitism and predation as being different forms of the same basic kind of interaction. In predation and almost all examples of parasitism one organism eats another organism. In predation, the predator usually kills its prey before it eats it. In parasitism, the **parasite** usually lives on or in a much larger organism and feeds on it while it is still alive. The parasite's unlucky "partner" is called its **host.** Parasites usually do not kill their host, although many weaken it greatly. Why do you think it is an advantage to a parasite not to kill its host?

A few parasites do kill their host. For example, certain wasps lay their eggs on caterpillars. The eggs hatch into wormlike young wasps that burrow into the body of the caterpillar. The young wasps feed on the caterpillar's tissues, avoiding the caterpillar's major organs so that it stays alive. After about 30 or 40 days, the young wasps chew their way out of their dying host's body and spin cocoons. Inside the cocoons, the young wasps develop into adult wasps.

Some organisms are considered parasites even though they do not feed on their host. Cuckoos (and a few other types of birds) lay their eggs in other birds' nests. When the baby cuckoo hatches, it pushes the eggs and young of its foster parents out of the nest. The foster parents are tricked into feeding and caring for the intruder as if it were their own offspring. How does the cuckoo's behavior harm other birds? Why are cuckoos considered to be parasites?

**Figure 1–20** *The vampire bat bites host animals with its razor-sharp fangs, then drinks their blood (bottom right). The white ovals on this caterpillar are the eggs of a wasp (top right). When the young wasps hatch, they will slowly eat the caterpillar. The cuckoo continues to trick its unlucky foster parents into caring for it—even after it has grown much larger than they are (top left). Why are these animals considered to be parasites?*

## Adapting to the Environment

Have you ever heard the phrase "survival of the fittest"? Hiding behind this simple phrase is a complicated process. In response to the challenges of their environment, species evolve, or change over time. The evolutionary changes that make organisms better suited for their environment occur by means of a process known as natural selection. Natural selection works like this: Only the individuals that are best suited for their environment survive and produce offspring. These offspring inherit the characteristics that made their parents well suited for the environment. Over the course of a number of generations, these well-suited individuals continue to thrive and reproduce. At the same time, the characteristics that make individuals poorly suited for the environment disappear. This is because the individuals that have these characteristics are less likely to survive. Such individuals have few, if any, offspring to inherit their characteristics. The net result of natural selection are changes in the behavior and physical characteristics of species that make them better suited for their environment. This process is called adaptation.

**Figure 1–21** *One adaptation to the environment involves the size of an animal's ears. Small ears, such as those of the arctic fox, help to conserve body heat (top). Large ears, such as those of the desert fox, help to get rid of excess heat (bottom). How else are these foxes adapted to their physical environment?*

Organisms cannot choose how they change. They also cannot invent new characteristics. Even with these limitations, living things have been changed in many strange and marvelous ways through natural selection to meet the challenges of their environment. Some of the most interesting adaptations have their origin in the way organisms interact with other living things in their community.

**ADAPTING TO PREDATORS** Organisms have evolved many adaptations to defend themselves against predators. Animals, such as deer and rabbits, are able to escape from predators by running very fast. Turtles, snails, and coconuts have hard shells that shield them from attackers. Skunks, stinkbugs, and mustard plants produce odors that repel predators. Fawns (baby deer) and many insects have colors and shapes that allow them to blend into their backgrounds. So do the flounder fish you read about earlier. Wasps, sea anemones, and nettle plants can sting. Roses, porcupine fish, hedgehogs, and sea urchins have long sharp thorns or spines. Toads,

puffer fish, certain mushrooms, and many plants contain poisonous chemicals. And seventeen-year cicadas and the seeds of century bamboo plants show up so rarely that predators are not used to eating them. (Imagine waiting a hundred years—or even seventeen years—for your next meal!)

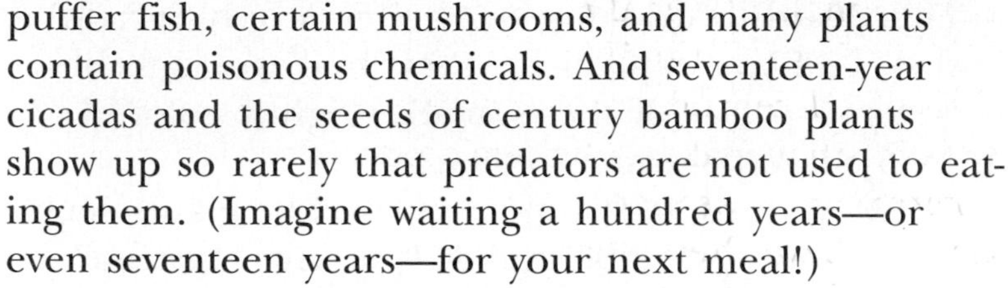

**Figure 1–22** *Some organisms defend themselves by looking like something they're not. This caterpillar looks like a bird-eating viper (right). What do the spiny bugs (left) and the praying mantis (center) resemble?*

But predators are not so easily discouraged! They have evolved in ways that help them get around their prey's defenses. Cheetahs, for example, can achieve bursts of speed that allow them to catch swiftly running gazelles.

Some types of animals—prairie dogs, musk oxen, bees, and ants, to name a few—live in large groups. Individuals in these groups can take turns looking out for enemies. They can also band together to fight off predators. Can you think of other ways in which living in groups helps organisms to survive? Why can living in groups be considered an adaptation?

**ADAPTING TO COMPETITORS** Organisms have evolved many adaptations for dealing with competition. One strategy is for competing species to divide up the habitat. Take a look at the Central and South American hummingbirds in Figure 1–24 on page 34. What do you notice about them?

Although these birds are similar in form, it is obvious that they have very different beaks. Each different (specialized) beak limits a particular

**Figure 1–23** *Brilliant colors warn predators that the sea slug is too dangerous to eat. The orange projections on the sea slug's back are loaded with stinging structures from the sea anemones it eats (top). What adaptations has the cactus evolved that protect it from predators?*

**Figure 1–24** *The long-tailed hermit (left), sicklebill (center), and green violet-ear (right), Central and South American hummingbirds, feed on the nectar of flowers. What is the most obvious way in which these birds have evolved to avoid competition?*

hummingbird to drinking nectar from only certain flowers. Can you explain how this makes it possible for several species of these small, brightly colored birds to share the same habitat?

A different sort of strategy is used by plants such as sunflowers, mesquite, and purple sage. Instead of sharing their habitat, these plants reduce competition by killing off their competitors. They release chemicals into their surroundings that discourage the growth of other plants.

The bacterium *Streptomyces* and the fungus *Penicillium* also produce chemicals that discourage the growth of competitors. You may be familiar with these chemicals in a more common form—that of antibiotics. Antibiotics help humans cure diseases by killing the microorganisms that cause them. The drug streptomycin is obtained from *Streptomyces.* Penicillin is obtained from *Penicillium.*

**ADAPTING IN SYMBIOSIS** The partners in many symbioses are extremely well-adapted to each other. For example, certain acacia trees in Latin America and Africa have huge hollow thorns or hollow swellings the size of Ping-Pong balls on their branches. These structures are inhabited by ants. The acacias also have other types of structures on or near their leaves that produce food for the ants. Thus the acacias feed the ants as well as provide them with a home. In return, the ants catch and eat small herbivores, such as grasshoppers, that land on their tree. If a large herbivore tries to nibble on the acacia's leaves (or if a curious person touches the acacia), the ants swarm onto the intruder, biting and stinging. The ants that live on the Latin American acacias also help their partner deal with competitors. They chew off any tree branches that come into contact with their acacia. This enables the tree to get plenty of light even though it lives in a dense tropical forest. Imagine the series of evolutionary adjustments that were necessary to make the ants and acacias so well suited to living together!

**Figure 1–25** *The round structures on the whistling thorn acacia provide a home for symbiotic ants.*

## 1–3 Section Review

1. Describe the three basic types of symbiosis. Give two examples of each.
2. What is competition? Why does it occur?

**Critical Thinking—*Making Generalizations***

3. The myxomatosis virus is spread by mosquitoes. When the virus was first introduced to Australia, it killed many rabbits very quickly. Now, the virus is slower and less deadly. Explain how and why the rabbits and virus have changed. What impact might this have on the Australian ecosystem?

# 1–4 Life in the Balance

**Guide for Reading**

*Focus on this question as you read.*

- *What are some ways in which humans can affect the balance of an ecosystem?*

Earlier in this chapter you learned that all of the living and nonliving things in an environment are interconnected, like the strands of a spider's web. Touching a single strand can cause the entire web to tremble. If too many threads are broken, the web collapses and must be built anew.

An ecosystem, however, is not an unchanging structure like a spider web. Changes are constantly occurring in an ecosystem. Populations increase and decrease. Trees fall and animals die. The weather changes with the seasons. Birds fly to warmer places for the winter. Each time a change occurs, an adjustment in the balance of an ecosystem is required.

Sometimes ecosystems are thrown completely out of balance by a natural disaster, such as a hurricane, landslide, forest fire, or volcanic eruption. Perhaps

**Figure 1–26** *In May 1980, the lush green forests of Mount St. Helens were destroyed by a volcanic eruption. The blast knocked over trees and covered the surrounding area in volcanic ash. But eventually life began to return to the area. The ecosystem of Mount St. Helens is slowly regaining its balance.*

you have read about or possibly even experienced the explosive eruption of Mount St. Helens in 1980, the forest fires in Yellowstone National Park in 1988, or Hurricane Hugo in 1989. What effects did these disasters have on the ecosystems in the area? In certain cases, however, events that seem like natural disasters are not really disasters at all. Instead, they are events that may actually help to maintain the balance in certain ecosystems. For example, naturally occurring forest and brush fires may clear away shrubs and dead wood, creating room for new plants to grow. Burning also breaks down dead plant materials and thus returns nutrients to the soil. Some organisms in fire-prone areas have evolved in ways that enable them to cope with periodic fires. A few even need fires. The seeds of the jack pine tree, for example, require the temperatures of a forest fire in order to be released from their protective pine cones.

Ecosystems are also put out of balance by human activities. **The damaging effects of activities such as chopping down forests or putting poisonous chemicals into rivers are quite obvious. Apparently harmless human actions may also cause widespread damage.** One of the more dramatic examples of the unexpected ways in which human activities can damage ecosystems involves Mono Lake, in eastern California.

Mono Lake is a beautiful saltwater lake fed by streams of melting snow from the Sierra Nevada Mountains. Two small islands within Mono Lake attract thousands of birds, especially seagulls. More than 80 species of birds nest on these islands and feed on the lake's shrimp, flies, and algae. That is, they did until 1981. During the summer of that year, many baby seagulls were found dead. How strange this seemed to be for an ecosystem that had long provided food, water, and shelter for seagulls! As people investigated the situation, they discovered that the ecosystem had been disturbed by actions taken far away from the lake many years before.

About 40 years ago, the city of Los Angeles began to use water from the major streams that feed into Mono Lake. As less and less water emptied into Mono Lake, the lake began to dry up. Thousands of acres of dust formed where there was once water.

As the amount of water in the lake decreased, the concentration of salt dissolved in the water increased. The shrimp that seagulls fed on could not survive in water so salty. As the shrimp died, less food was available for the seagulls. Baby seagulls starved to death.

To make matters worse, as the water level in Mono Lake dropped, a land bridge that connected the shore to the nesting islands formed. Coyotes crossed this bridge, killed many seagulls, and invaded the gulls' nests.

**Figure 1–27** *In January 1991, the Iraqi army occupying Kuwait dumped millions of liters of oil into the Persian Gulf, causing one of the world's worst environmental disasters. What effect did this oil spill have on wildlife?*

**Figure 1–28** *Mono Lake has the potential to be home to thousands of birds (left). But when too much of its water supply is taken away, the lake dries up and becomes virtually lifeless (right).*

Many people want to save Mono Lake. But it will not be easy. In order to conserve water and save the lake, some profitable farmland must be allowed to turn back into desert. In addition, people must change their lifestyles and give up nice things like green lawns, swimming pools, decorative fountains, and weekly car washes. As you can imagine, the government of California is going to have a hard time getting people to cooperate with water-saving measures. To complicate the situation, the area around Los Angeles received less rainfall in the late 1980s and early 1990s than it usually does. This has made the water shortage even more severe.

Without a water conservation plan, Mono Lake will continue to dry up. And the delicate balance between living and nonliving things in this ecosystem may be altered beyond all hope for recovery.

The story of Mono Lake is only one example of the effects humans have on the balance in ecosystems. There are thousands and thousands of other stories. Some end sadly with something beautiful or strange lost forever. A few have happy endings. And many are unfolding even as you read these words. With careful planning, people can help these stories end happily. By understanding the interactions within the environment, people can both use the resources around them and preserve the beauty, diversity, and balance within ecosystems.

### ACTIVITY WRITING

*Who Missed the Moose?*

Isle Royale is a long, narrow island in Lake Superior, 25 kilometers from the shore of Canada. In the early 1900s, a few moose swam to the island. Within 20 years, the population had increased to more than 2000!

During the next 40 years, the moose population underwent several changes. Using books and other materials in the library, find out what changes occurred on Isle Royale and what caused the changes. Include answers to the following questions in your report.

**1.** What are limiting factors and in what ways do they affect a population?

**2.** How do birth and death rates affect a population?

**3.** What is a population cycle?

## 1–4 Section Review

1. How can human activities affect the balance of ecosystems?
2. How did humans change the balance in the Mono Lake ecosystem?

**Connection—*You and Your World***

3. A picture-perfect lawn or golf course is usually a lush green expanse of just one kind of grass. Its appearance is kept up by a program of constant mowing, fertilizing, watering, and applying pesticides and weedkillers. Using what you have learned about interactions and balances in an ecosystem, explain why "perfect" lawns and golf courses are so difficult to maintain.

# CONNECTIONS

## *Down for the Count*

You might recall watching one of the adults in your household filling out a *census* form in 1990. Every 10 years, the government conducts a census of the people of the United States. Special forms are mailed out to every household in the country. People report the number of people in their household, the sex and age of each person, their income, the number of rooms in their dwelling, and many other facts. Then they mail back the forms.

Occasionally, there are problems. Some people give silly answers. Others fail to return their forms. And a few people may be overlooked by the census takers.

These problems seem minor compared to those faced by biologists performing a wildlife census. For one thing, wild animals do not have a mailing address! And even if they did, they could not fill out a census form! So the biologists must travel to the places where the animals live and put up with insect bites, awful weather, and other discomforts.

Some large, conspicuous organisms—such as caribou, elephants, and wildebeests—are simply counted from aerial photographs. The sizes of the populations of most other organisms are determined indirectly: Biologists count or estimate the size of part of a population, then use this figure to calculate the probable size of the entire population. For example, a biologist studying wolves may let out a howl, then count the number of answering howls from real wolves. Suppose the biologist knows that about a quarter of the wolves in the area will respond to the fake howl. If the biologist hears eight responding howls, how many wolves are probably in the area?

Some wildlife census methods involve catching animals. As you can imagine, this can be more hazardous than ringing doorbells in a city or town for the Census Bureau! While counting animals, biologists have been bitten, scratched, pecked, sprayed by skunks, menaced by bears, and even chased up a tree by a moose that recovered too soon from a dose of tranquilizer delivered by a dart gun.

Faced with all these difficulties, why do biologists conduct a census of wildlife? Because knowing the size of populations and how populations have changed over time gives people some of the information they need for deciding how to manage wildlife and the environment.

# Laboratory Investigation

## A Little Off Balance

### Problem

How does adding lawn fertilizer affect the balance of an aquatic (water) ecosystem?

### Materials *(per group)*

2 2-L wide-mouthed jars
pond water
8 *Elodea* (or other aquatic plant)
lawn fertilizer (or house plant food)
teaspoon

### Procedure

1. Label the jars A and B.
2. Fill each jar about three-fourths full with pond water.
3. Place four *Elodea* in each jar.
4. Add one-half teaspoon of lawn fertilizer to jar B.
5. Place the jars next to each other in a lighted area.
6. Predict what will happen to the jars over the course of three weeks. Record your predictions.
7. Observe the jars daily for three weeks. Record your observations.

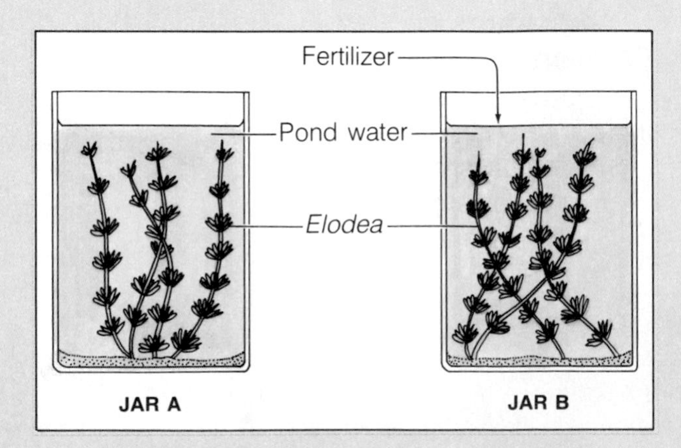

### Observations

1. Were there any differences between jars A and B? If so, what were the differences? When did you observe them?
2. How did your results compare to your predictions?

### Analysis and Conclusions

1. What was the control in this experiment? The variable?
2. Why did you place the jars next to each other? Why did you place them in the light?
3. What effect did the fertilizer have on the *Elodea*?
4. Lawn fertilizer contains nitrogen, phosphorus, and potassium. These nutrients are often present in sewage as well. Predict the effects of dumping untreated sewage into ponds and lakes.
5. **On Your Own** Design an experiment to test the effects of different amounts of lawn fertilizer on the balance of an aquatic ecosystem. Predict the results of your experiment. If you receive the proper permission, you may perform your experiment and find out if your predictions are correct.

## Summarizing Key Concepts

### 1–1 Living Things and Their Environment

▲ All of the living and nonliving things in an environment are interconnected.

▲ An ecosystem consists of all the living and nonliving things in a given area that interact with one another.

### 1–2 Food and Energy in the Environment

▲ Producers are the source of all the food in an ecosystem.

▲ Consumers cannot make their own food. They feed directly or indirectly on producers.

▲ Decomposers break down dead organisms into simpler substances. In the process, they return important materials to the soil and water.

▲ A food chain represents a series of events in which food and energy are transferred from one organism in a ecosystem to another.

▲ A food web consists of many overlapping food chains.

▲ The amount of energy at each feeding level in an ecosystem can be diagrammed as a pyramid of energy.

### 1–3 Interaction and Evolution

▲ An organism's niche consists of everything the organism does and everything the organism needs in its environment.

▲ Two or more species cannot share the same niche.

▲ Interactions such as predation, competition, and symbiosis have had a powerful effect on the course of evolution.

### 1–4 Life in the Balance

▲ Ecosystems are sometimes thrown completely out of balance by natural disasters or by human activities.

## Reviewing Key Terms

*Define each term in a complete sentence.*

### 1–1 Living Things and Their Environment

environment
ecology
ecosystem
community
population
habitat

### 1–2 Food and Energy in the Environment

producer
consumer
decomposer
food chain
food web

### 1–3 Interaction and Evolution

niche
competition
predator
prey
symbiosis
commensalism
mutualism
parasitism
parasite
host

# Chapter Review

## Content Review

### Multiple Choice

*Choose the letter of the answer that best completes each statement.*

**1.** The study of the interactions between living things and their environment is called
a. commensalism. c. ecology.
b. parasitism. d. botany.

**2.** Evolutionary changes usually occur by means of a process called
a. mutualism. c. predation.
b. natural selection. d. competition.

**3.** A desert is an example of a(an)
a. ecosystem. c. food chain.
b. population. d. niche.

**4.** A group of organisms of the same species living together in the same area is called a(an)
a. population. c. community.
b. ecosystem. d. niche.

**5.** Which term best describes the relationship between a honeybee and a flower?
a. commensalism c. competition
b. predation d. mutualism

**6.** An organism that eats plants is best described as a(an)
a. omnivore. c. scavenger.
b. herbivore. d. carnivore.

**7.** Everything an organism does and needs in its environment is known as its
a. feeding level. c. habitat.
b. niche. d. adaptation.

**8.** Which term best describes the relationship between a fox and a wolf?
a. competition c. predation
b. symbiosis d. mutualism

### True or False

*If the statement is true, write "true." If it is false, change the underlined word or words to make the statement true.*

**1.** Producers break down dead organisms.

**2.** Any close relationship between two organisms in which one organism lives near, on, or even inside another organism and in which at least one organism benefits is known as mutualism.

**3.** The living part of an ecosystem is called a community.

**4.** The organisms on which a predator feeds are known as its hosts.

**5.** An animal that feeds on the bodies of dead animals is known as a scavenger.

**6.** The relationship between a dog and a flea is an example of commensalism.

**7.** Consumers, such as green plants, are the first link in a food chain.

**8.** Within a food chain, there is more energy available at each higher feeding level.

### Concept Mapping

*Complete the following concept map for Section 1–1. Refer to pages G6–G7 to construct a concept map for the entire chapter.*

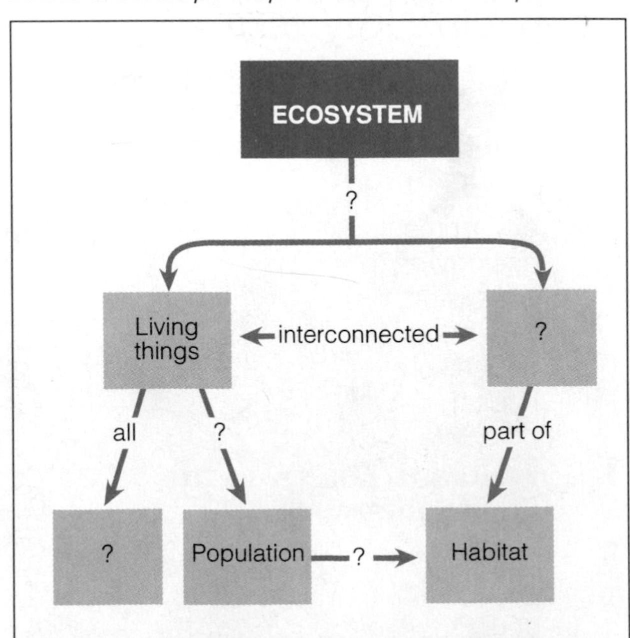

## Concept Mastery

*Discuss each of the following in a brief paragraph.*

1. Explain how the terms environment and ecosystem differ in meaning.
2. Name and briefly describe the three basic energy roles in an ecosystem. For each role, give an example of an organism that plays that role.
3. Explain how the following terms are related to one another: community, ecosystem, feeding level, habitat, niche, population, species.
4. How are a food chain, a food web, and a pyramid of energy different from one another? Describe the relationship among the three.
5. Using specific examples, explain how predation, competition, and symbiosis can affect evolution.
6. Explain this statement. "In the environment, change is a two-way street."

## Critical Thinking and Problem Solving

*Use the skills you have developed in this chapter to answer each of the following.*

1. **Applying concepts** African tickbirds can often be found perching on large animals such as Cape buffalo and rhinoceroses. The tickbirds eat bloodsucking ticks found on the skin of the large animals. What type of symbiosis is this? Explain.
2. **Making diagrams** Draw a food web that includes the following organisms: cat, caterpillar, corn, cow, crow, deer, hawk, human, lettuce, mouse, fox, grass, grasshopper, rabbit. Identify each organism as a producer or a consumer.

3. **Making predictions** How could the spraying of an insecticide interfere with the balance in an ecosystem?
4. **Relating concepts** Explain why the sun is considered to be the ultimate source of energy for almost all ecosystems.
5. **Identifying relationships** Take a look at the pyramid of energy in Figure 1–13 on page 22. What trend would you expect to see in the number of organisms at each level as you move from the bottom of the pyramid toward the top? Why?
6. **Assessing concepts** "If an ecosystem is properly selected, it contains four basic parts: the physical environment, the living things, energy, and the nutrients that circulate between its living and nonliving elements." Examine the accompanying photograph of a deer licking a block of salt. Explain whether you can identify each of these elements in the photograph. Then explain whether you think this description of an ecosystem is a useful one.
7. **Using the writing process** Write a short story or poem about a change in the balance of an ecosystem that you yourself have observed.

# Cycles in Nature

## Guide for Reading

*After you read the following sections, you will be able to*

**2–1 Cycles in Time: Rhythms of Life**

- Describe how biological clocks affect organisms.
- Give examples of the ways in which the rhythms of life are linked to cycles in time.

**2–2 Cycles of Matter**

- Discuss how matter flows through an ecosystem.

**2–3 Cycles of Change: Ecological Succession**

- Describe how ecosystems are changed by the process of succession.

On a cool spring night in southern California, the ocean waves surge up onto the beach. As each wave retreats, small silver fish known as grunion (GROON-yuhn) appear. The wet sand near the top of the high tidemark glitters with grunion.

The female grunion squirm into the sand tail first. They then deposit their pale orange eggs below the surface of the sand. The male grunion curve their bodies around their mates and fertilize the eggs. Then the next wave washes over the beach and sweeps the grunion back to sea.

Grunion deposit their eggs at places on the beach reached only once every two weeks by the highest tides. Hidden beneath the sand, the eggs are safe from the waves—and egg-eating predators in the ocean. By the time the tides are at their highest point again and the waves reach the eggs, the young grunion are ready to hatch. As the waves swirl the eggs from the sand, the tiny grunion pop out and are carried to the ocean. The process repeats itself year after year, generation after generation. It is an age-old cycle that ties the survival of a tiny fish to the movements of the sea, the Earth, and the moon. It is one of the many cycles in nature. Turn the page, and learn about more.

### Journal *Activity*

***You and Your World*** What is your favorite season? Why? What observations have you made about nature as the seasons change? Explore your thoughts and feelings in your journal.

*On spring and summer nights with a full moon or a new moon, grunion ride the ocean waves onto the beaches where they mate and lay their eggs.*

**Guide for Reading**

*Focus on these questions as you read.*

- *What controls the rhythms of life?*
- *What are some examples of daily, lunar, and annual rhythms?*

# 2–1 Cycles in Time: Rhythms of Life

What do you think of when you hear the word rhythm? A musician might think of the beat of a song or a dance. A soldier might think of the pace of marching in a parade. A biologist might think of the way a heart beats, a seagull flaps its wings, or a cricket chirps. Or a biologist might think of slower rhythms than these—for example, the way fiddler crabs change their color from light gray to dark gray and back again to light gray during the course of a day. Or the way whales travel from cold polar waters to warmer regions closer to the equator during the course of a year. A rhythm is any pattern that occurs over and over again.

Slower biological rhythms are often in harmony with certain natural cycles, such as the passage of day into night, the rise and fall of the tides, and the changing seasons. As summer changes to autumn, for example, the leaves of maple trees become red, yellow, and orange in color.

One of the most interesting things about many of the slower rhythms of life is that they continue even if an organism is removed from its natural environment. Once a year, ground squirrels will go into their winter sleep and starlings (small black birds) will breed as if it were spring—even when they live in a cage in the unchanging, seasonless world of a laboratory. Human volunteers living in a sunless cave

**Figure 2–1** *The elegant flowers of the night-blooming cereus open in the evening and close in the morning. In certain areas, the leaves of many kinds of trees change color during the autumn. Once a year, albatrosses woo a mate by strutting about and bobbing their heads. Why are these events considered to be examples of biological rhythms?*

without any way to tell time still experience daily changes in blood pressure, body temperature, wakefulness, and other biological functions. Evidence from experiments such as these indicate that many of life's rhythms are not simply responses to changes in the environment. Something inside humans, squirrels, birds, and other living things keeps track of the passage of time. But what exactly is this something that monitors the passage of time?

Internal timers known as **biological clocks** may be responsible for keeping track of many different cycles of time. These cycles may range in length from a few minutes to many years. When the time is right, biological clocks "tell" organisms to change their appearance, behavior, or body functions in some way. For example, biological clocks tell grunion when to ride the waves onto a beach. **Biological clocks help living things stay in step with rhythmic cycles of change in their environment.**

To better understand why biological clocks are so important to living things, it might help you to think about the way alarm clocks help people stay in step with their daily activities. Have you ever used an alarm clock to help you get up in the morning? If you have, you know that it would be silly to set your alarm clock for the time that you want to leave for school. You need time to get out of bed, get dressed, eat breakfast, and so on. So if you need to be at school at 8:00, you might set your alarm clock for 6:45. This gives you time to get ready for your day. Biological clocks are important to living things for the same reason—although they may measure

**Figure 2–2** *Biological clocks help living things stay in step with their environment. Biological clocks let snow geese know when it is time to fly south for the winter and tell morning glories to open their flowers during the day.*

## CAREERS

### Forester

**Foresters** carefully study a possible recreation area to be sure that such items as cooking equipment, picnic tables, utility sources, and roads do not disturb the ecological balance of the forest. Other activities of foresters range from managing forest timberland to clearing away tree branches so that sunlight can reach the ground cover.

Those interested in pursuing a forestry career can get a start by studying chemistry, physics, math, earth science, and biological science in high school. Summer job experience in forest and conservation work is also helpful. Forestry applicants should enjoy working outdoors, be physically fit, and work well with people. For further information, write to the Society of American Foresters, Wild Acres, 5400 Grosvenor Lane, Bethesda, MD 20814.

time periods other than 24-hour days! How might having biological clocks be better for organisms than simply responding to changes in the environment as they happen?

In nature, biological clocks are extraordinarily accurate. However, under unchanging conditions (in a laboratory, for example), biological clocks usually run a little too slow or a little too fast. For example, in a laboratory setting where light, temperature, and all other factors remain the same, a fiddler crab's cycle of color changes might take 23 hours rather than 24. Each day, the crab would change color an hour earlier than crabs in their natural environment. After a while, the crab's internal cycle would be completely out of step with its natural environment.

Why don't organisms get out of step in nature? The answer is simple: Biological clocks are set and reset by environmental cues such as dawn or dusk, day length, moisture, and temperature. But because biological clocks are influenced by the environment, it is not always easy to tell which changes in organisms are caused by a biological clock and which are caused by environmental cues. It is easy to see, however, that the rhythms of life are linked to natural cycles in time. In the next few pages, you will read about some of the ways the rhythms of life are in step with daily, lunar (of the moon), and annual (yearly) cycles.

## Daily Rhythms

As night falls, the creatures of the day prepare to sleep. A flock of birds circles high above some trees, then suddenly drops into the branches. Colorful flowers close. A dog turns around several times before curling up on its bed with a sigh. Meanwhile, the creatures of the night are becoming active. A swarm of bats bursts from a cave. Mushrooms emerge and grow among the dead leaves on a forest floor. Certain microorganisms in the ocean begin to glow with an eerie bluish light.

Organisms that are active during the day are said to be **diurnal** (digh-ER-nuhl). Those that are active at night are said to be **nocturnal** (nahk-TER-nuhl). Are humans diurnal or nocturnal?

Evolution has shaped the characteristics of organisms in such a way that diurnal organisms are well suited for the warm, dry, brightly lit day and nocturnal organisms are well suited for the cool, moist, dimly lit night. Take a look at the nocturnal night monkey and the diurnal emperor tamarin in Figure 2–3. What is the most obvious difference in the facial features of these two monkeys (other than the emperor tamarin's moustache)? That's right—the "eyes" have it! Like many nocturnal animals, night monkeys have much larger eyes than their diurnal relatives. Larger eyes gather a larger amount of the available light and allow the nocturnal animals to find their way in the darkness of the night.

Not all nocturnal animals have oversized eyes. Many rely on other senses to guide them in the dark. For example, owls and bats rely a great deal on their sense of hearing. And many noctural insects have extremely long feelers that allow them to explore the nighttime world through their senses of touch, taste, and smell.

**Figure 2–3** *The nocturnal owl monkey (bottom) and the diurnal emperor tamarin (top right) both live in the forests of Latin America. How might different patterns of wakefulness in these monkeys help to reduce competition between them? Some organisms are neither nocturnal nor diurnal. The vole (top left) is most active at dawn and at dusk.*

## Lunar Rhythms

Have you ever spent an entire day at the beach? If so, you might have noticed that each successive wave seemed to reach a little less far up the beach. After a while, you might have realized that the level of the ocean had dropped, revealing once-hidden rocks, seaweed, barnacles, and mussels.

**Activity Bank**

Going Through a Phase, p.142

**Figure 2–4** *Organisms that live at the edge of the ocean—such as the giant green sea anemone (bottom right), fiddler crab (top right), ringed top snail (top left), and red East African starfish (bottom left)—show patterns of activity that match the rise and fall of the tides.*

An examination of the rocks, wet sand, and tide pools at the water's edge might have revealed some interesting creatures becoming active at the low tide. Certain crabs pop out of their burrows and scurry around on the wet sand. Small starfish push their stomachs out of their mouths and absorb bits of food from the surface of the sand. Microscopic one-celled diatoms and small green worms rise to the surface of the sand, creating small, faint patches of brown and green.

Other organisms become inactive at low tide. Mussels and barnacles shut their shells tightly. Snails and other small animals hide beneath seaweed or within cracks in rocks. Sea anemones pull in their tentacles and contract into sand-covered blobs.

If you stayed at the beach long enough, you might have seen the waves come in once more, slowly washing away sand castles and gradually hiding the things that you had seen at low tide. With the return of the water at high tide, the crabs retreat to their burrows and block the entrances with sand. The diatoms and worms sink back under the sand to avoid being washed away by the waves. Snails begin to creep around. Mussels and barnacles open their

shells and filter food from the water. And sea anemones open up in flowerlike splendor.

The rise and fall of the tides are controlled by the moon (and to a lesser degree, the sun). As a result, tidal rhythms are considered to be lunar rhythms. (The word lunar means of the moon.) There are two kinds of tidal rhythms. You have just read about the high- and low-tide cycle, which occurs twice a day (every 12.4 hours, to be exact). The other type of tidal rhythm is roughly a two-week cycle (14.8 days) during which the high tides gradually become higher, peak, and then decrease.

Biological clocks in a number of organisms are in harmony with the two-week cycle of the tides. The grunion you read about at the beginning of the chapter are an example. They lay their eggs on the nights of the highest high tide. On a small island west of Australia, female red crabs release their eggs into the ocean during the lowest high tide. In tropical oceans, strange worms break in two at night during the lowest high tide. One half remains in its burrow at the bottom of the sea. The other half swims to the surface. There it joins millions of other half-worms. Many of these swimming half-worms are eaten by fishes, birds, and humans (who consider the worms a treat!). The remaining half-worms burst open at sunrise, releasing huge numbers of eggs and sperm.

As you might expect, 12-hour and 2-week tidal rhythms occur mostly in organisms that live near and in the ocean. However, some land organisms have rhythms that are in harmony with the lunar cycle. For example, the average length of a human pregnancy is exactly ten lunar months from the time the egg cell is released to the time the baby is born.

## Annual Rhythms

Some of life's rhythms are closely associated with the seasons of the year. In the spring, for example, songbirds build their nests. Bears awaken from their winter sleep. Trees that were bare throughout the winter begin to grow a new cover of leaves. Daffodils, tulips, and primroses bloom. And many animals give birth to their young. Can you think of some other events that occur in the spring? How

*But What Will You Do When It Grows Up?*

This is a question often asked of someone who is raising a wild animal. It is also a question people must ask themselves as their "pet" progresses in its life cycle. Some books that deal with this question are: *Rascal,* by Sterling North; *The Yearling,* by Marjorie Rawlings; *The Story of Elsa,* by Joy Adamson; and *Ring of Bright Water,* by Gavin Maxwell.

**Figure 2–5** *Chrysanthemums normally bloom in the autumn in response to lengthening nights. How do florists cause chrysanthemums to produce flowers for Mother's Day?*

# ACTIVITY WRITING

*All the Right Moves*

Write a report on the annual migrations of the animal of your choice. Some migratory animals include: springbok, monarch butterflies, ruby-throated hummingbirds, storks, certain eels, sea turtles, and certain bats. In your report, include a map that shows the route of the animal's migration.

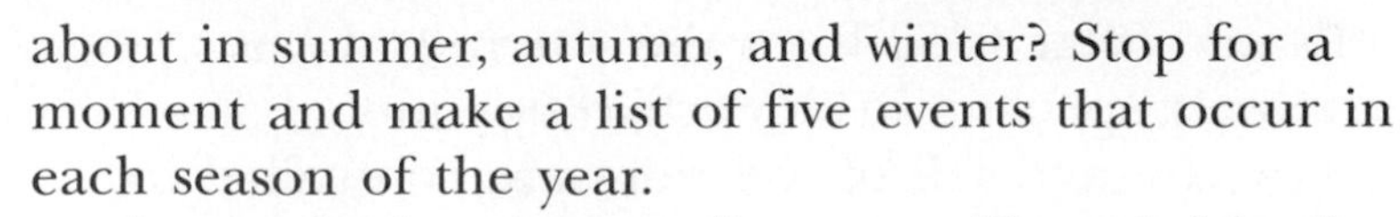

about in summer, autumn, and winter? Stop for a moment and make a list of five events that occur in each season of the year.

As you look over your list, you will probably discover that many of the events occur once a year, every year. Such events are examples of yearly, or annual, rhythms.

There are many examples of annual rhythms in nature. Many animals—such as grunion, deer, and red crabs—reproduce only during certain times of the year, when conditions are best for the young to survive. Can you explain why most animals bear their young in the spring, rather than in the winter?

In response to seasonal variations in temperature, food supply, light, and other factors, some organisms migrate. This means they journey from one place to another. For example, many birds in the northern hemisphere migrate south for the winter. In the spring, the birds migrate back to northern regions to breed and raise their young. Birds are not the only organisms that migrate, however. Wildebeest, whales, bats, salmon, and red crabs are just a few of the other living things that migrate.

For the most part, **migrations** are annual rhythms in which organisms travel from the place where they breed to the place where they feed. In general, organisms migrate to more beneficial environments as seasonal changes make their old environment less favorable.

**Figure 2–6** *Caribou migrate between their summer home in the far north and their winter home in the forests hundreds of kilometers to the south. During their annual migration on Christmas Island, red crabs appear everywhere!*

Some organisms have a different way of escaping unfavorable seasonal changes. They simply "sleep" through the bad periods of the year. These organisms are active only when conditions are favorable. For example, as winter approaches, the body functions of toads, bears, and certain other animals slow down. This enables these animals to wait out the cold winter months in sheltered hiding places. In their slowed-down state, the animals can survive without food or water until the coming of spring. This winter resting state is known as **hibernation.** (The Latin word *hibernus* means winter.)

In places where the summer months are extremely harsh, hot, and dry, organisms may enter a resting state known as **estivation.** (The Latin word *aestas* means summer.) For example, as the shallow lakes in which African lungfish live begin to dry up, the lungfish may bury themselves in the mud at the bottom of the lake. They can survive for many years in their shell of dried mud.

A few organisms, such as certain plants and insects, simply do not live through the harsh seasons in their environment. But before they die, these organisms produce weather-resistant eggs, spores, or seeds. The eggs, spores, or seeds are able to grow and develop when the seasons change and conditions become favorable once more. The life cycle of such organisms is usually an annual (yearly) cycle. For example, the plants known as annuals—such as marigolds, petunias, and sweet peas—start from seeds in spring, mature and bear seeds by autumn, and die in winter. The next spring, new plants sprout from the seeds and the cycle continues.

**Figure 2–7** *A chipmunk escapes the cold, hungry days of winter by hibernating. A spadefoot toad avoids the harshest conditions of its desert home by estivating in an underground burrow.*

**Figure 2–8** *Annual organisms, such as California poppies, simply do not live through the most unfavorable seasons. How do annual organisms ensure that their species continue from year to year?*

## 2–1 Section Review

1. What is a biological clock? How do biological clocks affect organisms? Give three examples of events controlled by a biological clock.
2. How are hibernation and estivation similar? How are they different?
3. What is migration? How does migration help organisms survive?

**Critical Thinking—*Applying Concepts***

4. A person who has just traveled a long distance by airplane (for example, from New York to Hawaii) might start to fall asleep during dinner and be wide awake at three in the morning. This condition is sometimes known as "jet lag." Explain why jet lag occurs.

**Guide for Reading**

*Focus on this question as you read.*

- *How do water, oxygen, carbon, and nitrogen flow through the environment?*

# 2–2 Cycles of Matter

There are many types of cycles in nature. The changing seasons form a cycle, as do the rising and falling tides and the passage of day into night. Organisms have life cycles in which they are born, grow, reproduce to create the next generation, and eventually die. Stars also undergo a series of changes from the time they are "born" as hot spinning clouds of gases to the time they "die" explosively. Rocks are worn down into sand, which may then be transformed by heat and pressure back into rock. Circular series of chemical reactions, or chemical cycles, turn carbon dioxide gas into food and break down food to release energy and carbon dioxide.

In Chapter 1, you learned about the way energy flows through an ecosystem. As you may recall, energy is used up at each feeding level. But ecosystems constantly receive a new supply of energy—usually in the form of sunlight.

**Figure 2–9** *Cycles in nature do not always involve living things. New stars may be formed from the matter that remains after a star explodes.*

**Figure 2–10** *In the rock cycle, sand may be transformed into rock, which may then be worn down to sand. The slow wearing-down of sandstone may produce arches and other amazing rock formations.*

**Activity Bank**

A Saucepan Simulation of a Cycle, p.144

The supply of matter in an ecosystem, however, is not renewed. But matter, unlike energy, can be recycled, or reused. **Matter, in the form of chemicals, flows in cycles from the nonliving part of the environment to living things and back again.**

There are many cycles of matter, and most of them are quite complex. Fortunately, you do not need to know every detail of every cycle. In this section, you will learn about the basic steps (and only the basic steps!) of four of the most important cycles of matter: the **water cycle,** the **oxygen cycle,** the **carbon cycle,** and the **nitrogen cycle.**

## The Water Cycle

For many of the world's cultures, water has long symbolized life. Countries with water usually prospered—and those that lacked water often faced disaster. Knowing where water was and where it would be was a key to success. Many thousands of years ago, people became aware that there is a natural cycle to the flow of water on this planet—a cycle people still rely on today.

"Earth" is not a particularly appropriate name for our planet. "Water" would actually be more fitting, for three fourths of the planet Earth is covered by lakes, streams, rivers, and oceans. Water circulates continuously between the Earth's surface and the atmosphere (the envelope of air surrounding the Earth). Water on the Earth's surface is heated by the sun and evaporates. In other words, it changes from a liquid into a gas, or vapor. The water vapor then

**Figure 2–11** *From a distance, the Earth is blue with large bodies of water and white with clouds. In Chinese mythology, dragons were in charge of the Earth's rivers, bodies of water, and rain. Why is this dragon surrounded by tiny clouds?*

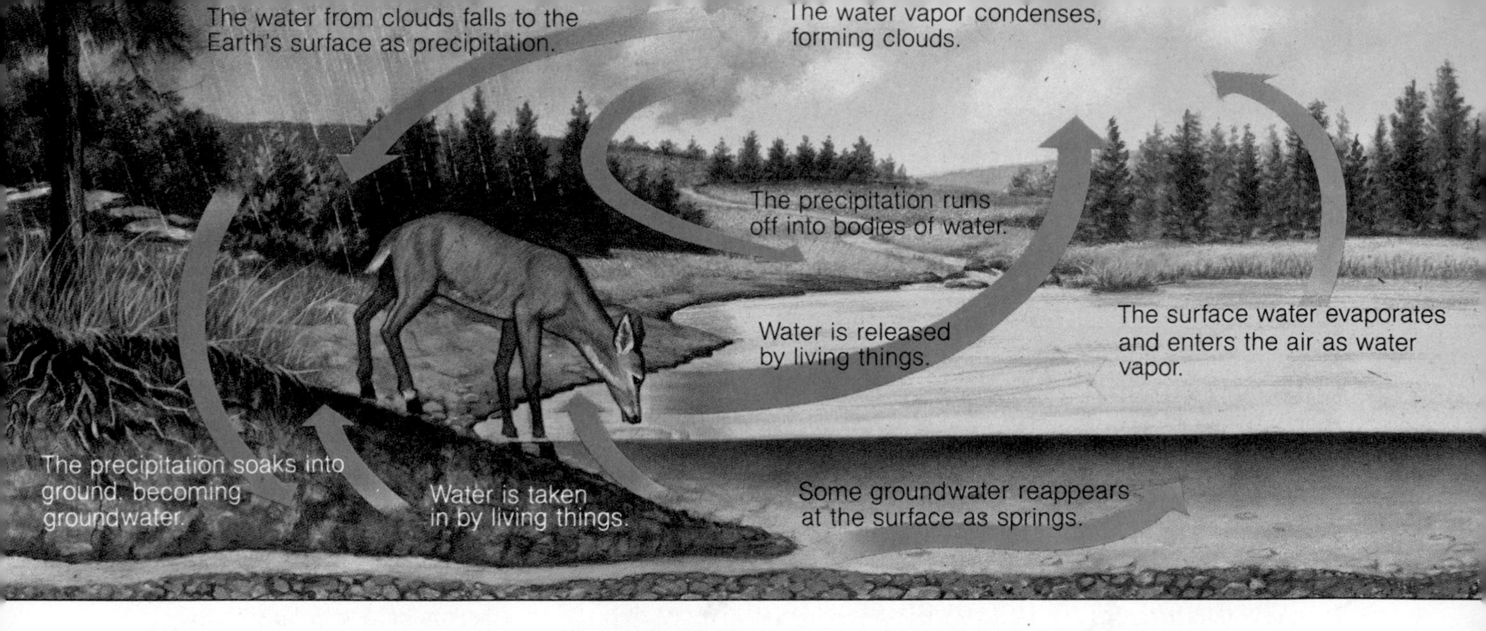


**Figure 2–12** *Water cycles through both the living and the nonliving part of the environment. What happens to water that falls as precipitation?*

Take It With a Grain of Salt, p.146

*It Takes Your Breath Away*

Hold a mirror a few centimeters from your mouth. Say "Horace the horse hulas in Hilo, Hawaii." What happens to the mirror? Why does this happen?

■ How does this relate to the water cycle?

rises up into the air. In the upper atmosphere, water vapor cools and condenses into liquid droplets. It is these droplets that form clouds. Eventually, the droplets fall back to the surface of the Earth as precipitation—rain, snow, sleet, or hail.

Most precipitation falls directly back into the oceans, lakes, rivers, and streams. Some of the rest falls on the surface of the land and then runs off into these bodies of water. In either case, water that evaporates into the air returns to the surface of the Earth, and the cycle repeats itself.

Not all water, of course, goes directly back into the Earth's bodies of water. Some is taken in by living things and later returned to the nonliving part of the environment. For example, plants take in liquid water through their roots and release some water vapor through their leaves. Animals drink water, but they also give water back to the environment when they breathe and in their wastes.

## The Oxygen and Carbon Cycles

Like most living things, you need oxygen to survive. The atmosphere, which is 20 percent oxygen, supplies you and other air-breathing organisms with this vital gas. Oxygen from the atmosphere that has dissolved in water is breathed by fish and other aquatic organisms. Clearly, living things would have used up the available oxygen supply in the atmosphere millions of years ago if something did not return the oxygen to the air. But what could that something be?

Consider this: When you inhale, you take in oxygen. When you exhale, you release the waste gas carbon dioxide. If something used carbon dioxide and released oxygen, it would balance your use of oxygen. That something, as you may already know, are producers such as green plants and certain microorganisms. These producers use carbon dioxide gas, water, and the energy of sunlight to make carbon-containing compounds that are often referred to as "food." During the food-making process, the producers also produce oxygen, which is released into the environment. Through this process, known as the oxygen cycle, there is always a plentiful supply of oxygen available for air-breathing organisms.

But what happens to the carbon in food? How is it transformed back into carbon dioxide? In order to extract energy from food, organisms must digest the food, or break it down into simpler substances. This process ultimately produces water and carbon dioxide, which are released back into the environment. Figure 2–14 illustrates the oxygen and carbon cycles. Can you explain why these two cycles are usually discussed together?

**Figure 2–13** *The water in this tiger's breath is visible as white droplets condensing in the cold winter air. What other substances does the tiger exhale?*

1. Carbon dioxide is used by producers such as green plants.
2. Oxygen is released by producers such as green plants.
3. Oxygen is used by air-breathing organisms.
4. Carbon dioxide is released by air-breathing organisms.

**Figure 2–14** *Carbon and oxygen cycle between producers and other living things, including elephants. How does the carbon in carbon dioxide become the carbon in food?*

## The Nitrogen Cycle

About 78 percent of the atmosphere is made up of "free" nitrogen, or nitrogen that is not combined with other elements. All living things need nitrogen to build proteins and certain other body chemicals. However, most organisms—including plants, animals, and fungi—cannot get the nitrogen they need from the free nitrogen in the air. They can use only nitrogen that is combined with other elements in compounds. But how are these nitrogen-containing compounds made?

Certain kinds of bacteria are able to use the free nitrogen in the air to make nitrogen compounds through a process known as nitrogen fixation. Most of the nitrogen fixation on Earth occurs as a result of the activity of bacteria. Some of these bacteria live

*Miniature Worlds*

Certain companies make sealed glass globes that contain air, water, some small water plants, a few tiny shrimp, and certain microorganisms. If kept in a sunny place, the living things in the glass globe can survive indefinitely. How do the nitrogen, oxygen, and carbon cycles work in these miniature ecosystems?

in the soil. Others live in the water. Others grow inside special structures on the roots of certain plants, including beans, clover, alfalfa, peas, and peanuts.

One family of nitrogen compounds produced by nitrogen-fixing bacteria consists of substances called nitrates. Nitrates can be taken from the soil by plants. Inside the plants, the nitrogen in the nitrates is used to make compounds such as proteins. The compounds made by the plants can be used by animals, fungi, and other organisms that cannot use nitrates directly. Take a moment now to look at Figure 2–15. Trace the steps of the nitrogen cycle from the free nitrogen in the air to the nitrogen in the bodies of animals.

You have just read about the part of the nitrogen cycle in which nitrogen is transferred from the nonliving portion of the environment into living things. Now let's look at the part of the nitrogen cycle that returns nitrogen to nonliving things.

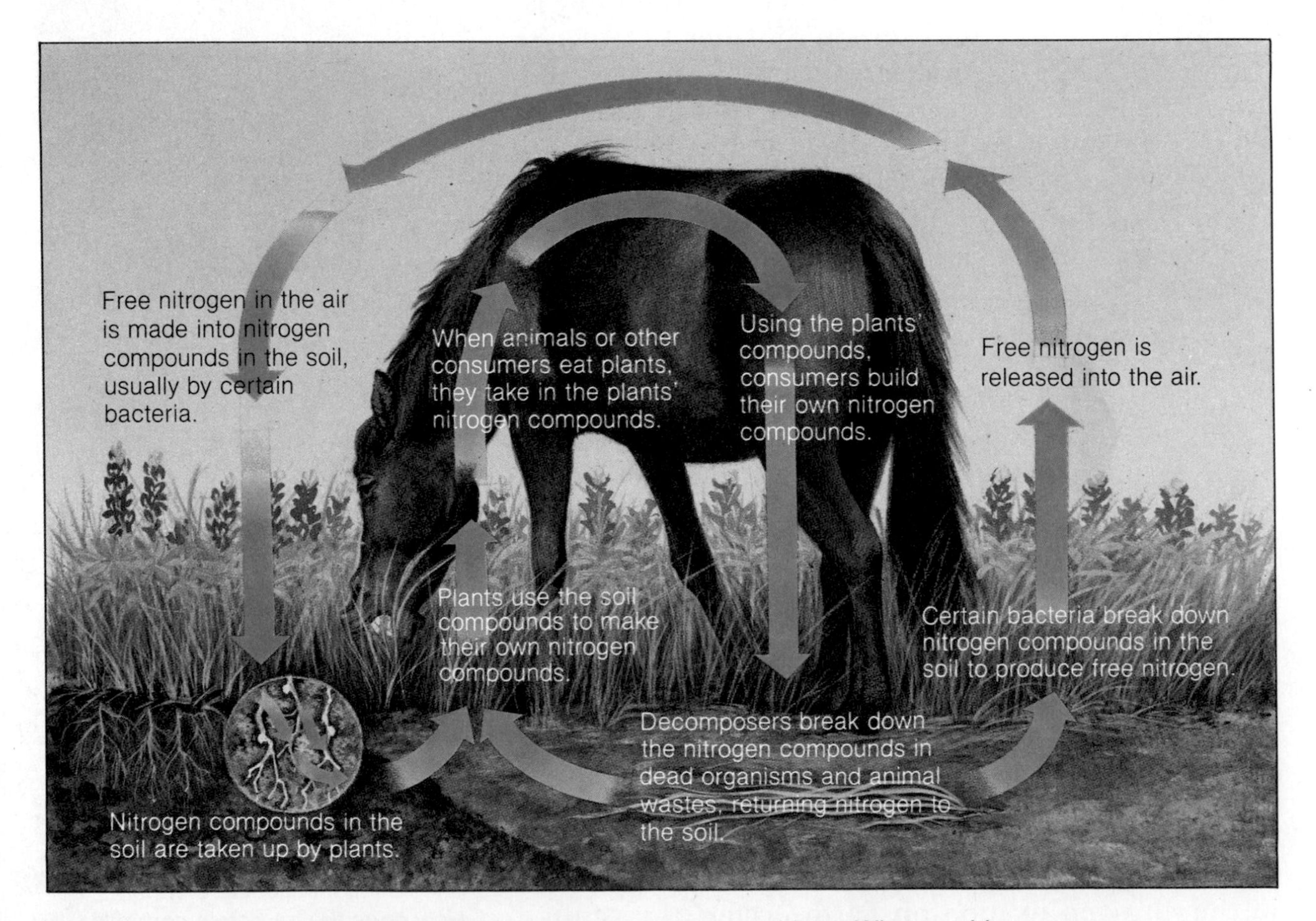


**Figure 2–15** *Nitrogen flows through the air, the soil, and living things. What would happen to the nitrogen cycle if all bacteria suddenly vanished from the Earth?*

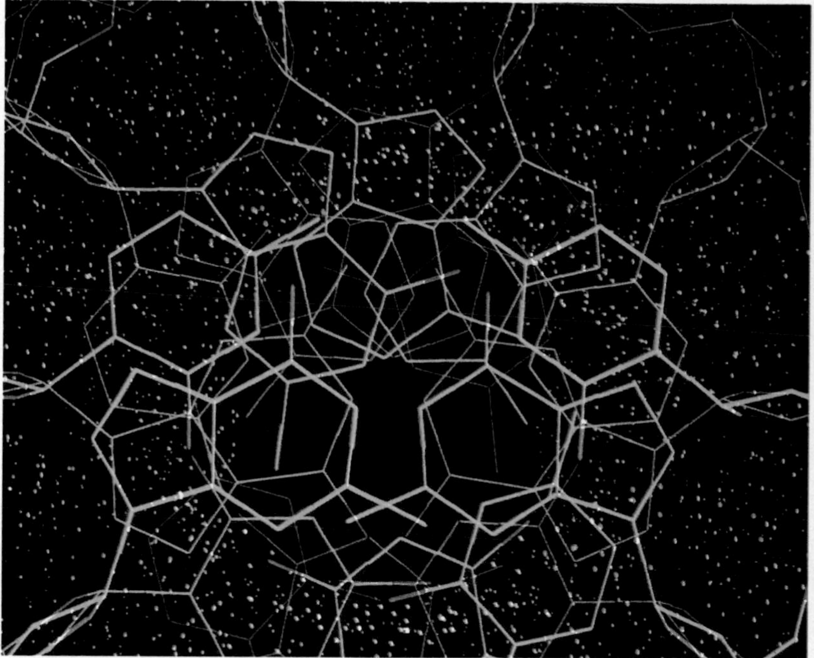

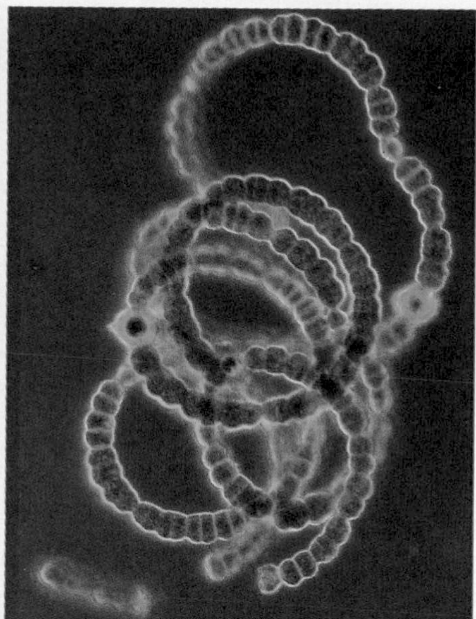

Decomposers, such as certain bacteria, break down the complex nitrogen compounds in dead organisms and animal wastes. This returns simple nitrogen compounds to the soil. These simple compounds may be used by bacteria to make nitrates.

Nitrogen can go back and forth between the soil and plants and animals many times. Eventually, however, certain kinds of bacteria break down nitrogen compounds to produce free nitrogen. The free nitrogen is released into the air, completing the cycle.

**Figure 2–16** *Lightning may cause chemical reactions that change free nitrogen into nitrogen compounds. Most nitrogen fixation on Earth, however, is performed by bacteria (top right). The fixed nitrogen can then be used to produce proteins and other important compounds, including DNA (top left).*

## 2–2 Section Review

1. Show the basic steps of the nitrogen, water, oxygen, and carbon cycles by drawing a simple diagram for each cycle. (Do not copy the diagrams in your textbook!)
2. Why is it important for matter to be recycled in ecosystems?

**Connection—*Ecology***

3. Every fourth year, a farmer plants alfalfa or clover on a field instead of wheat. Explain this practice.

**Guide for Reading**

*Focus on this question as you read.*

- ▶ *What is ecological succession?*

# 2–3 Cycles of Change: Ecological Succession

Imagine that you have built a time machine in a secret clearing in a forest. The big day has arrived—you are finally ready to test your invention! Your hands shake with excitement as you set the controls. You are about to travel hundreds of years back in time. You start the machine, and . . .

Splash! You find yourself in the middle of a pond. What's going on here? You started in a forest and now you're in a pond. Can a forest have once been a pond? You decide to use your time machine to discover an answer to this question. (Fortunately, your time machine is waterproof!) On your way back to the future, you will make a few stops so you can look around and see if your surroundings change over time.

At first, the pond is quite deep. No plants grow at the very bottom of its center. Not enough light can penetrate through the deep water for plants to survive. The pond is inhabited by fishes, the young of insects such as dragonflies, and a huge number of small aquatic animals and microorganisms.

As time passes, particles of dirt, fallen leaves, and the remains of dead water organisms begin to accumulate at the edges and bottom of the pond. As the pond becomes more shallow, new organisms can get a foothold. Underwater plants line the bottom of the pond. Eventually, water plants that poke out of the water—such as water lilies, reeds, and cattails—start growing around the edges of the pond.

**Figure 2–17** *The process of succession is considered to begin with bare rock, such as that resulting from a lava flow, or with a newly formed pond. Succession then gradually changes the area. How did succession affect this pond up to now? How will it probably change the pond in the future?*

As the pond continues to fill in, the fish begin to die off. They are replaced by air-breathing animals such as frogs and turtles. The water lilies, reeds, and other plants grow all across the pond. Materials from the dead plants further fill in the pond.

Eventually only a few patches of open water are left. The pond has become a marsh. Like the pond, the marsh keeps filling in. In time, it becomes dry land. Rabbits and deer roam where fish and frogs once lived. Bushes and then trees take root. What began as a pond has become a forest.

Over time, one ecological community succeeds, or follows, another. **The process in which the community in a particular place is gradually replaced by another community is called ecological succession.** And, as your time-machine adventure has shown you, the process of **ecological succession** can completely change what a place looks like.

If the community in a particular place is left alone, it may in time consist of a group of species that are not replaced by new arrivals. This stable collection of plants, animals, and other organisms is known as a **climax community.** The climax community varies from place to place. In the northeastern United States, for example, a climax community may be characterized by oak and hickory trees. In certain areas of northern California, the climax community may be dominated by huge redwood trees.

The process of succession does not have to start with a pond or other body of water. It can begin with the bare rock formed by a lava flow or landslide. It can also occur on soil that has been cleared of vegetation by a disaster or in areas where some sort of ecological disturbance such as logging or farming has been stopped.

Succession usually takes a long time. To go from a tiny pond to a forest can take more than a hundred years. Outside forces, however, can affect the rate of succession. For example, certain pollutants, such as sewage and phosphate-containing detergents, can cause the plants in a body of water to grow extremely quickly. How would this affect succession?

**Figure 2–18** *Succession resumes when an area, such as a corn field, is left alone. The first year, a few weeds and grasses take root. After two years, the field is covered with grass. After ten years, there are shrubs and young trees. After twenty years, the field has become a young forest.*

**Figure 2–19** *In 1988, terrible fires destroyed much of the forest in Yellowstone National Park. How did this affect succession?*

Some events can slow down succession or set it back a few steps. Plowing a field prevents bushes and trees from gaining a foothold. Fire can burn a developing forest and set it back. Floods can fill a dying pond with water again. Because of these kinds of changes in the environment, succession usually moves in cycles rather than in a straight line.

When making decisions about how natural resources should be used, it is important to keep this in mind: Succession can take many different paths and can lead to different climax communities. The end result of succession is due in a large part to chance. A particular pond, for example, may end up as a forest of willow and alder trees or a treeless peat bog. Because of this, people cannot assume that an ecosystem will recover and return to the way it was after it has been disturbed. It is possible that succession will restore an ecosystem after trees are cut down, land is dug up for mining, oil is spilled, or any other kind of environmental damage occurs. But it is also possible that succession will take a different path and the ecosystem will never be the same again. Because of this, people must be very careful in deciding which resources should be used. Succession will occur—but no one can accurately predict the course it will take.

## ACTIVITY
### DISCOVERING

*Backyard Succession*

**1.** Locate a backyard, field, or park lawn that is usually mowed.

**2.** After obtaining the proper permission, measure an area that is 1 meter on each side.

**3.** Do not mow or disturb that area for six weeks or longer.

**4.** After six weeks, carefully explore your area to see what organisms are living in it.

■ Compare the organisms you find in your area with the ones that are growing in the mowed area. What conclusion can you reach?

## 2–3 Section Review

1. What is ecological succession? How do ecosystems change as a result of succession?
2. What is a climax community?

**Critical Thinking—*Appraising Conclusions***

3. Logging companies often plant pine seedlings in places where they have cut down all the trees. They argue that there is nothing wrong with harvesting trees from any forest, as long as measures such as replanting are taken. Many ecologists argue that forests that have not yet been affected by logging should be left alone. Using what you have learned about succession, explain why the ecologists take this position. Do you think this is reasonable?

# CONNECTIONS

## Cycles and Stories

Why are there seasons? What causes day to turn into night? Why does the moon seem to grow and then shrink during the course of a month?

Today we know that these never-ending cycles of change are caused by the movement of the Earth and the moon. But long ago, people did not know about the way the Earth and moon spin through space. As they looked in wonder at the world around them, people created stories, or *myths* (MIHTHS), to explain what they saw. Here is one such myth from Nigeria, a country in Africa.

### Why the Moon Grows and Shrinks

The bush baby (a small monkeylike animal) was very poor. His friend the mouse-deer felt sorry for him, and wanted to help. So the mouse-deer, who had two pairs of eyes, gave one pair to the bush baby.

The mouse-deer's eyes were large, round precious stones that shone with a light of their own. But no one could afford to buy such fine gems. So the bush baby broke the eyes into tiny pieces.

Unfortunately, when the bush baby went to sell the tiny sparkling gems, the wind blew them all over the town. It took the children of the town a month to gather up the tiny gems and put them in a box. But as soon as they had finished, the wind began to blow them out of the box once more.

Every month, the wind scatters the bush-baby's tiny gems across the town of the sky, where they glitter as stars. And each month, the children gather up the stars and put them into the box of the moon. But as soon as the moon is full, the wind begins to blow away the gathered gems.

■ What are some myths about nature that belong to your own cultural heritage? Discover the myths of your ancestors by talking to the older members of your family or by going to the library.

# Laboratory Investigation

## Going in Cycles

### Problem

What are the steps in the life cycle of a housefly?

### Materials *(per group)*

magnifying glass
rubber band
cotton ball
glass jar
piece of gauze cloth large enough to cover the top of the glass jar
20 mL bran flakes
10 mL diluted canned milk
paper towel
houseflies
metal bottle cap

### Procedure

1. To make a fly cage, place a paper towel in the bottom of a glass jar. Put 20 mL of bran flakes and 10 mL of diluted canned milk on the towel.
2. Wet a cotton ball with water and put it in the metal bottle cap. Put the bottle cap, with the cotton ball facing up, into the glass jar.
3. Put the flies your teacher gives you into the cage. Stretch the gauze cloth over the mouth of the jar and hold it in place with the rubber band.
4. Using the magnifying glass, check to be sure there is at least one female fly. Female flies have pointed tails and small eyes. Males have rounded tails and large eyes.
5. When eggs appear, release your adult flies outdoors. Fly eggs are small, white, and shaped like sausages.

### Observations

1. Each day, check the jar and write a description of what you see.
2. For how many days do you see eggs?
3. Larvae (wormlike young flies) come from the eggs. Draw a larva. For how many days do you see larvae?
4. A larva becomes a pupa. Draw a pupa. For how many days do you see pupae?
5. What does a pupa become?

### Analysis and Conclusions

1. Using your drawings and other data from your investigation, prepare a diagram showing the life cycle of a housefly.
2. **On Your Own** Design an investigation to examine the life cycle of another kind of organism—a bean plant, fruit fly, guppy (a fish), or frog, for example. If you receive the proper permission, you may perform the investigation you have designed.

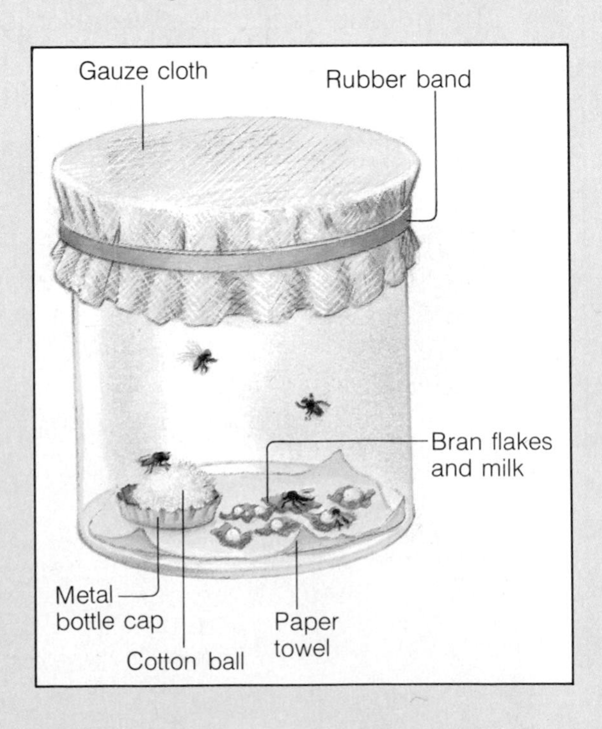

## Summarizing Key Concepts

### 2–1 Cycles in Time: Rhythms of Life

▲ Biological clocks help living things stay in step with rhythmic cycles of change in their environment.

▲ Biological clocks are set and reset by environmental cues such as dawn or dusk, day length, moisture, and temperature.

▲ The rhythms of life are linked to daily, lunar, and annual cycles in time.

▲ Organisms that are active during the day are said to be diurnal. Those that are active at night are said to be nocturnal.

▲ Tidal rhythms are one type of lunar rhythm. There are two kinds of tidal rhythms: a roughly two-week cycle and a roughly 12-hour cycle.

▲ Events that occur once a year, every year are examples of annual rhythms.

▲ Migration is the movement of organisms from one place to another. Many animals have annual migrations.

▲ In winter, some organisms enter a resting state known as hibernation.

▲ In summer, some organisms enter a resting state known as estivation.

### 2–2 Cycles of Matter

▲ Unlike energy, matter can be recycled. Matter flows in cycles from the nonliving part of the environment to living things and back again.

▲ There are many cycles of matter in ecosystems. Four of the most important are the water, oxygen, carbon, and nitrogen cycles.

### 2–3 Cycles of Change: Ecological Succession

▲ The process in which the set of living things in a particular place is gradually replaced by another set of living things is called ecological succession.

▲ In time, a particular place may possess a stable collection of plants, animals, and other organisms known as a climax community. The climax community varies from place to place.

▲ Succession usually takes a long time. Outside forces, however, can affect the rate of succession and may even reset the cycles of succession.

## Reviewing Key Terms

*Define each term in a complete sentence.*

### 2–1 Cycles in Time: Rhythms of Life

biological clock
diurnal
nocturnal
migration
hibernation
estivation

### 2–2 Cycles of Matter

water cycle
oxygen cycle
carbon cycle
nitrogen cycle

### 2–3 Cycles of Change: Ecological Succession

ecological succession
climax community

# Chapter Review

## Content Review

### Multiple Choice

*Choose the letter of the answer that best completes each statement.*

1. Organisms that are active during the day are said to be
   a. nocturnal. c. lunar.
   b. diurnal. d. annual.
2. In the winter, frogs and ground squirrels enter a resting state known as
   a. succession. c. hibernation.
   b. estivation. d. migration.
3. Fiddler crabs are most active at low tide. What kind of rhythm are the fiddler crabs showing?
   a. annual c. daily
   b. diurnal d. lunar
4. The act of traveling to a new environment when seasonal changes make the old environment less favorable is known as
   a. succession. c. hibernation.
   b. estivation. d. migration.
5. Clouds are formed from water vapor by
   a. condensation. c. precipitation.
   b. evaporation. d. denitrification.
6. Organisms that are active at night are said to be
   a. nocturnal. c. lunar.
   b. diurnal. d. annual.
7. Succession may result in a stable set of organisms known as a(an)
   a. ecosystem.
   b. climax community.
   c. transitional community.
   d. migration.
8. Most nitrogen fixation on Earth occurs through the activity of
   a. plants. c. animals.
   b. bacteria. d. fungi.

### True or False

*If the statement is true, write "true." If it is false, change the underlined word or words to make the statement true.*

1. Biological clocks are inner timers that help organisms stay in step with natural cycles in time.
2. Unlike energy, matter cannot be recycled.
3. The movement of organisms from the place where they feed to the place where they breed is known as estivation.
4. The approximate 12-hour cycle of the tides is an example of a daily rhythm.
5. Liquid water changes into water vapor through precipitation.
6. Succession can be speeded up, slowed down, or reset by outside forces.
7. Air-breathing organisms exhale carbon dioxide, which can then be used by producers such as green plants.

### Concept Mapping

*Complete the following concept map for Section 2–1. Refer to pages G6–G7 to construct a concept map for the entire chapter.*

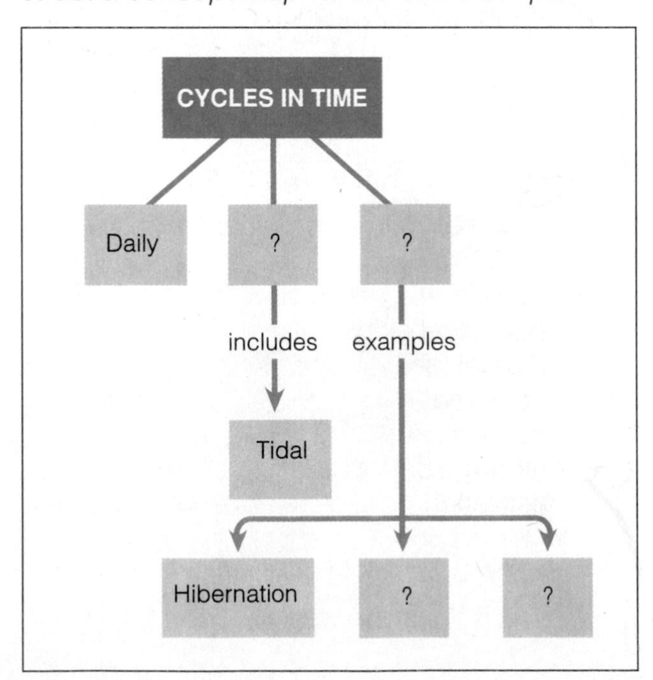

## Concept Mastery

*Discuss each of the following in a brief paragraph.*

1. Using specific examples, describe four different strategies for dealing with seasonal changes in the environment.
2. How is reproduction in grunion tied to daily, lunar, and annual cycles of time?
3. Do you live in a climax community? What observations lead you to this conclusion?
4. Why are cycles of matter important to living things?
5. How do the oxygen and carbon cycles link you to green plants?
6. Describe the basic steps of the nitrogen cycle.
7. What is succession? Explain how succession can change a marsh into a forest.

## Critical Thinking and Problem Solving

*Use the skills you have developed in this chapter to answer each of the following.*

1. **Relating cause and effect** There is much evidence to support the theory of the "greenhouse effect." According to this theory, excess amounts of carbon dioxide in the atmosphere can cause temperatures all over the world to rise. Explain how destruction of the world's forests may contribute to the greenhouse effect.
2. **Making predictions** A volcano is forming a new island in the ocean southeast of the island of Hawaii. This island will emerge from the sea in a thousand years or so. How might succession change this island over time? How might further volcanic eruptions on the island affect succession?
3. **Relating cause and effect** The accompanying photograph shows one use of fossil fuels—to power cars, buses, and trucks. As they burn, fossil fuels (coal, oil, natural gas, and gasoline, to name a few) release energy and carbon dioxide. For about two hundred years, people have been burning huge amounts of fossil fuels for energy. How does this affect the oxygen and carbon cycles?
4. **Making diagrams** Nitrogen in your food today may have once been part of a dinosaur. Draw a diagram that shows how the nitrogen might have gotten from the dinosaur to you.
5. **Assessing concepts** Is it better to think of succession as a one-way street or as a series of cycles? Explain your answer.
6. **Using the writing process** Imagine that you are a drop of water. Describe your journey through the water cycle. What changes do you undergo along the way? What living and nonliving things do you meet? What do you think about them? Is going through the water cycle fun or is it an unpleasant chore?

# Exploring Earth's Biomes

## Guide for Reading

*After you read the following sections, you will be able to*

**3–1 Biogeography**

- Explain how plants and animals move from one area to another.

**3–2 Tundra Biomes**

- Describe the characteristics of tundra biomes.

**3–3 Forest Biomes**

- Compare the characteristics of three forest biomes.

**3–4 Grassland Biomes**

- Describe the characteristics of grassland biomes.

**3–5 Desert Biomes**

- Describe the characteristics of desert biomes.

**3–6 Water Biomes**

- Identify and describe the characteristics of the major water biomes.

Night falls quickly on the vast Serengeti Plain of East Africa, as if a black velvet curtain has suddenly been drawn over the land. The scattered acacia trees and the great herds of zebras, wildebeests, and gazelles that graze on the Plain during the day disappear in the sudden darkness.

Safe in camp, you sit in your tent and listen to the mysterious sounds of the night. Nearby, a family of zebras snorts and stomps, startled perhaps by the rumble of distant thunder. The wildebeests, or gnus, stir and shuffle as they settle down for the night. And then the lions begin to roar. The wild music of the lions sends chills down your spine.

Lions, zebras, wildebeests, and gazelles do not live everywhere in Africa. They inhabit only the open plains, or savannas, with few or no trees and plenty of grass. Different animals live in the steamy jungles, which have many trees but not much grass. As you will discover in the pages that follow, animal and plant populations are not the same from place to place. They vary because different areas of the Earth have different climates. Climate conditions play a large role in determining where organisms make their homes.

### Journal *Activity*

***You and Your World*** Perhaps you have camped in a state or national park, spent the night in a tent in your own backyard, or imagined what it would be like to camp out. In your journal, describe your experiences, whether they are actual or imagined.

*Alert and watchful, a group of lionesses scan the grassy African plain for their prey.*

## Guide for Reading

*Focus on these questions as you read.*

- *How do plants and animals disperse?*
- *What are the six major land biomes?*

# 3–1 Biogeography

You are an explorer. In this chapter, you are going on a trip around the world. Your trip will take you from the cold, barren lands surrounding the North Pole to the dense jungles near the equator—and even into the depths of the oceans. As you travel, you will discover many strange and wonderful plants and animals. You will find that the kinds of plants and animals change as you move from place to place on your journey around the world.

The study of where plants and animals live throughout the world (their distribution) is called **biogeography.** Biogeographers, then, are interested in ecology, or the study of the relationships among plants, animals, and their environment.

The kinds of animals that live in an area depend largely on the kinds of plants that grow there. Do you know why? Animals rely on plants as one source of food. For example, zebras eat mostly grass. They would have a difficult time finding enough food in a jungle, where grass is scarce. But grassy plains are a good habitat, or living place, for zebras. Plains are also a good habitat for lions. Why? Lions are meat eaters (carnivores) that hunt plant eaters (herbivores), such as zebras, for food.

**Figure 3–1** *The gray-headed albatross makes its nest out of mud and grass on small wind-swept islands in the Southern Hemisphere (bottom left). Meerkats live in Africa's Kalahari desert (top). In what kind of African habitat would you expect to find a lowland gorilla (bottom right)?*

In turn, the plant life in an area is determined mainly by climate. Climate describes the average conditions of temperature and precipitation (rain, snow, sleet, hail) in an area over a long period of time. Trees grow tall and dense in warm, rainy climates, especially if the days are long and there is plenty of sunlight throughout the year. Fewer trees grow in cold, dry climates, where the short days of winter arrive early and stay late.

## Dispersal of Plants and Animals

In addition to studying where plants and animals live, biogeographers also study why plants and animals spread into different areas of the world. The movement of living things from one place to another is called **dispersal.** Plants and animals disperse in many ways. For example, about 50 million years ago, horses evolved in North America. During prehistoric times, the sea level dropped and a land bridge formed between Alaska and Siberia, which is in Asia. Horses soon moved westward across this natural land bridge into Asia. Over many thousands of years, horses dispersed all across northern Asia and into Europe.

Sometimes plants and animals disperse with help—from water, wind, and even people. Certain lizards, for example, have spread from island to island on floating branches. Some seeds, such as coconuts, also reach new places by floating on water. Certain microorganisms, the spores of fungi, dandelion seeds, baby spiders, and many other small, light organisms may be carried by the wind to new places.

**Figure 3–2** *Some organisms disperse with the help of water, wind, and other living things. The dispersal of coconuts (top) and lizards (bottom right) may be assisted by water. What helps the dispersal of dandelion seeds (bottom left)?*

## CAREERS

### Park Ranger

As you drive up to your campsite in Yosemite National Park, a **park ranger** is there to greet you. The ranger gives you a list of the guidelines you must follow while you are in the park. She also reminds you not to pick any plants growing in the park or feed any of the wild animals you might see around your camp.

Park rangers are responsible for enforcing the rules and regulations in state and national parks. They protect the plants and animals that live in the park and are also available to help visitors. A park ranger must have a college degree in forestry, park management, or some other related field.

If you are interested in ecology and enjoy working outdoors, you might want to consider a career as a park ranger. For more information, write to the National Park Service, United States Department of the Interior, Washington, DC 21240.

Often animals are brought to new homes by other animals. Fish eggs may be carried on the feet of ducks and other water birds. Insects may hitch a ride in the fur of mammals. You may be familiar with this form of dispersal if your pet dog or cat brings fleas into your home!

People have also been responsible for the dispersal of plants and animals. About one hundred years ago, a bird lover released some European birds called starlings in New York City. From New York, the starlings quickly spread across the country. Today, starlings are so common that some people consider them pests.

During the 1800s, ships bound for the Hawaiian Islands carried water for their crews in large barrels. Before the ships left their home ports, mosquitoes laid eggs in the water. The eggs hatched during the voyage. When the ships landed in Hawaii, they introduced mosquitoes to the islands. Unfortunately, the mosquitoes carried an organism that causes a serious bird disease called avian malaria. When the mosquitoes bit the Hawaiian birds, they transmitted the organism, causing the death of many birds.

Not all plants and animals carried to new homes by people are harmful. When European explorers came to the Americas they found the Native Americans growing corn, tomatoes, and squash. These plants were taken by the explorers to many parts of the world, where they are now important crops. Water buffaloes from southern Asia were brought to Europe and South America, where they became useful work animals.

## Barriers

After the prehistoric horses traveled from North America to Asia, the sea level rose again and covered the land bridge they had crossed. The sea became a natural fence, or barrier, that kept the horses from moving back and forth between the two continents. Eventually, horses became extinct (died out) in North America. Horses were unknown to Native Americans until European explorers arrived with them about 500 years ago.

Water is one of many natural barriers that can prevent plants and animals from dispersing. However, what may be a barrier for one kind of animal may not be for others. For example, water is a highway for fishes. Other natural barriers include deep valleys and high mountains.

Objects built by people may also be barriers. For example, suppose that a dam on a river acts as a barrier to salmon. Adult salmon cannot swim up the river to reach the places where they lay their eggs. Young salmon cannot reach the ocean, the place where they grow into adults. How do you think the dam will affect the salmon's ability to survive and reproduce? What will eventually happen to the salmon?

Natural barriers can also be ecological. This means that they have to do with an organism's relationship to its environment—both the living and the nonliving parts. When a habitat (living place) does not meet the needs of certain plants and animals, it is an ecological barrier. The Virginia opossum has spread from the South into the northeastern United States. During cold winters, opossums in northern states suffer from frostbite on their hairless ears and tails. It is a sign that they have met an ecological barrier—a cold climate—that probably will keep them from moving much farther north.

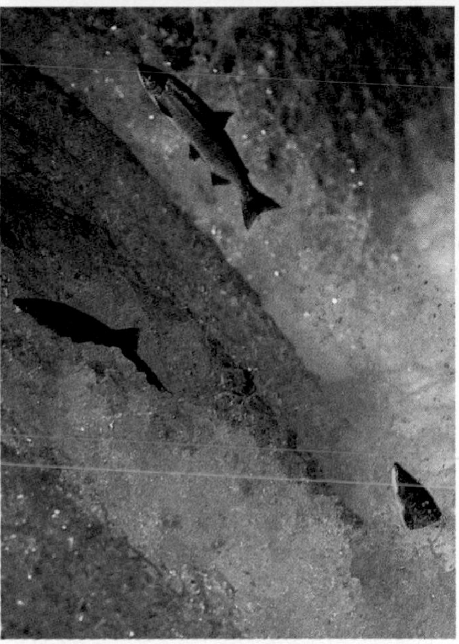

**Figure 3–3** *During their annual migration, salmon are able to swim and jump up small, natural waterfalls (bottom). But they cannot jump over artificial dams many meters high. Fish ladders—which look like large, low staircases covered by water—help the salmon get over dams (top).*

## Biomes of the World

The climate and the organisms living in an area give that area its special character. A grassland environment, for example, is quite different from a forest environment. Of course, sometimes it is a bit difficult to tell where one environment ends and another begins. In East Africa many grasslands are savannas—flat plains dotted with trees. When are there enough trees on a savanna to make it a forest and not a grassland? Settling such questions is a task for biogeographers.

To bring some order to the variety of environments on our planet, scientists have grouped environments with similar climates and ecological communities into divisions called **biomes.** Biome divisions are merely a classification system to help scientists describe the natural world. As you might expect, not all scientists divide the world into the same numbers and kinds of biomes. However, as a

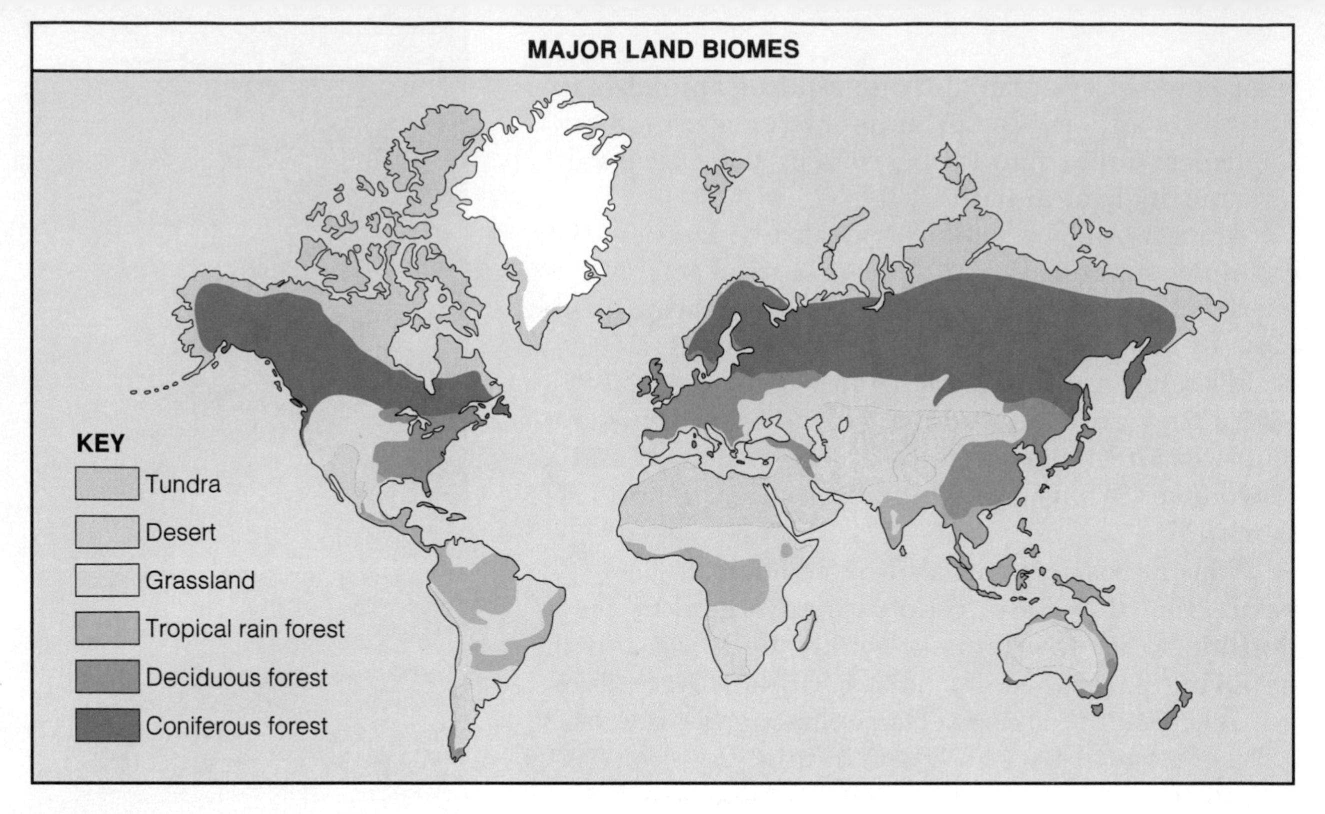


**Figure 3–4** *This map shows the distribution of biomes throughout the world. The white area on Greenland is an ice desert. In which biome do you live?*

rule, at least six land biomes are accepted by most scientists. **The six major land biomes are tundras, coniferous forests, deciduous forests, tropical rain forests, grasslands, and deserts.** In your exploration of the Earth's biomes, you will visit each one of these areas and discover something about the plants and animals that live there.

## 3–1 Section Review

1. Describe three ways in which plants and animals may disperse from one place to another. How do barriers prevent plants and animals from dispersing? Give two examples.
2. What are the six major land biomes?
3. How do scientists classify biomes?
4. What is biogeography?

**Critical Thinking—*Relating Concepts***

5. In some places where there are dams on a river, people have built structures called "fish ladders." How do you think a fish ladder helps the fishes get over a dam?

# 3–2 Tundra Biomes

**Guide for Reading**

*Focus on this question as you read.*

- *What is the climate of a tundra biome?*

The first stop on your journey through the Earth's biomes is the tundra. A tundra biome circles the Arctic Ocean all around the North Pole. You set up camp near the ocean in Canada's Northwest Territories. It is winter, and despite your heavy clothing, the wind cuts to the bone. **The climate of a tundra biome is very cold and dry.** A tundra is, in fact, like a cold desert. The temperature rarely rises above freezing (0°C). And during most years, less than 25 centimeters of rain and snow fall on the tundra.

Most water on the tundra is locked in ice within the soil. Even in spring and summer (which last a total of only three months!) the soil stays permanently frozen up to about a finger's length of the surface. The permanently frozen soil is called **permafrost.** Permafrost, along with the fierce tundra winds, prevents large trees from rooting. The few trees that do grow on the tundra are dwarf willows and birches less than knee high.

Among the most common tundra plants are lichens. Actually, lichens consist of fungi and algae growing together. Lichens cover the rocks and bare ground like a carpet. They are the main food of caribou, a type of reindeer. During winter, the caribou search out places where the snow is thinnest so they can find lichens easily. By the time the snow is

**Figure 3–5** *For part of the year, the tundra is dotted with shallow pools of water. The water cannot sink into the soil because of the permafrost layer. The permafrost is one reason tundra plants, such as the dwarf willow, do not grow very large. The thick, shaggy coats of musk oxen help them survive the long, cold tundra winters.*

**Figure 3–6** *The tundra birds called ptarmigans have white feathers in the winter. In the spring, the white feathers fall out and brownish feathers grow in. How does this change of feathers help ptarmigans survive?*

very deep, the caribou herds travel toward the forests south of the tundra. Wolves often follow close behind the herds, picking off the old and weak caribou.

As you leave your camp, you see great shaggy beasts with drooping horns pawing through the snow looking for dwarf willows to eat. They are musk oxen. Under their long outer coat is another coat of fine hair, which insulates them from the cold.

Many small animals inhabit the tundra too. Among the most common are lemmings, small rodents that look like field mice. When winter approaches, the claws of some kinds of lemmings grow thick and broad, helping them burrow in the snow, ice, and frozen soil. Lemmings spend the winter under the snow, feeding on green shoots and grasses.

You stay on the tundra until spring. As the surface soil melts, pools of water appear. Clouds of mosquitoes swarm around these pools. Unless you wear netting over your face, they make you miserable. With spring, the tough grasses and tiny flowers of the tundra burst into life. The sky is filled with birds. Vast flocks of ducks, geese, and shore birds, such as sandpipers, migrate from the south to nest on the tundra. Weasels and arctic foxes hunt the young birds in their nests. Ground squirrels, which hibernate in burrows during the cold winter months, awaken. The days are long and sunny, but some of the nights are frosty. The hint of frost warns you that on the tundra, winter is never very far off. It is time to continue your journey.

## 3–2 Section Review

1. Describe the tundra climate.
2. What is permafrost? What effect does permafrost have on the plant life of the tundra?
3. Unlike ground squirrels, lemmings do not hibernate during the winter. How do lemmings survive winter on the tundra?

**Critical Thinking—*Making Comparisons***

4. Why can a tundra biome be compared to a cold desert?

# 3–3 Forest Biomes

**Guide for Reading**

*Focus on this question as you read.*

- *What are the three major forest biomes?*

After leaving the tundra, you head south toward the Earth's forest biomes. **The three major forest biomes are coniferous forests, deciduous forests, and tropical rain forests.** Traveling south, you reach the coniferous forests first.

## Coniferous Forests

The northernmost forest biome, the coniferous forests, stretches in a belt across Canada, Alaska, northern Asia, and northern Europe. Fingers of these forests reach south along the high slopes of mountains such as the Rockies, where the climate is colder than in the lands below. Coniferous forests are made up of trees called **conifers.** Conifers, or evergreens, produce their seeds in cones.

Sometimes called "the great north woods," the coniferous forests have fewer types of trees than forests in warmer climates. Not many kinds of trees can stand the cold northern winters as well as firs, spruces, pines, and other conifers can. When you look closely at a conifer, you discover that its needles have a waxy covering. What purpose do you think this covering serves? You are right if you said it protects the needles from freezing. Because of the cold, fallen branches, needles, and dead animals do not decay as fast in coniferous forests as they do in warmer regions. Because the decay of plant and animal remains is one of the main factors in producing fertile soil, the soil of the coniferous forests is not particularly rich. Poor soil is another reason why many kinds of trees are unable to grow in coniferous forests.

Shade from the thick conifer branches, together with the poor soil, keeps many plants from growing on the forest floor. You find that as you hike through the north woods, you hardly ever have to hack through underbrush. Instead, the ground is usually covered with a thick, springy layer of fallen needles.

It happens to be late spring, however, so the going is not all that easy. The ground is spongy and soggy, and pools of water dot the forest floor. Unlike

**Figure 3–7** *Most of the trees in coniferous forests produce their seeds in cones.*

## ACTIVITY DOING

*Forest Food Webs*

On a piece of posterboard, construct a food web for each of the three forest biomes. Use string to show the connections between organisms. Label each organism. Identify the food chains in each biome.

permafrost, soil in a coniferous forest thaws completely in spring, making some parts of the forest like a swamp. Indeed, these areas are often called **taiga,** a Russian word meaning "swamp forest." The taiga includes not only these swampy areas but also the entire northernmost region of the forests.

Approaching a lake, you see a huge moose, shoulder deep in water. It dips its head and comes up with a mouthful of juicy water plants from the lake bottom. The lake, you discover, has been formed behind a dam of sticks and branches built by beavers across a stream. As you walk along the lake, you might just spot a Canadian lynx stalking a snowshoe hare. Or you might see one of the many members of the weasel family—perhaps a marten—hunting red squirrels. Some of the same kinds of animals you saw on the tundra also inhabit the coniferous forests—a fact you discover when you hear the howling of a wolf pack as you slip into your tent for the night.

Bird songs awaken you in the morning. Warblers, which leave for the south in autumn, twitter. Gray jays, which stay all year round, scold. A reddish bird with a crisscrossed bill lands on a pine tree branch just above your tent. It is a crossbill, a bird adapted to feeding on pine cones. In a moment, you see exactly how. The crossbill pries the scales of a pine cone apart with its bill and removes a seed with its tongue. Meanwhile, on the forest floor, another bird called a spruce grouse feeds on the needles and buds of spruce and other conifers. Packing your gear, you prepare to move on.

**Figure 3–8** *Animals that live in the coniferous forests of North America include moose, lynx, and beaver.*

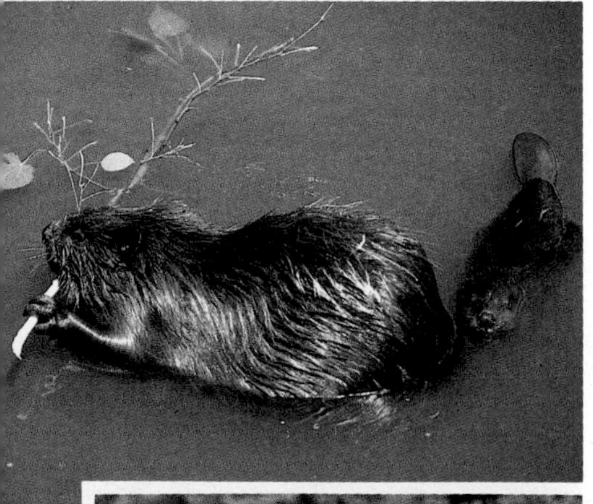

**Figure 3–9** *The great gray owl (left), crossbill (center), and spruce grouse (right) are just a few of the many birds that make their home in coniferous forests.*

## Deciduous Forests

Heading south from the coniferous forest, you reach a deciduous forest. Deciduous forests start around the border between the northeastern United States and Canada. They cover the eastern United States. Other deciduous forests grow throughout most of Europe and eastern Asia. Deciduous trees, such as oaks and maples, shed their leaves in autumn. New leaves grow back in the spring.

Deciduous forests grow where there is at least 75 centimeters of rain a year. Summers are warm and winters are cold, but not as cold as in the northern coniferous forests. You wander among oaks, maples, beeches, and hickories. A thick carpet of dead leaves rustles underfoot. The decaying leaves help make the soil of a deciduous forest richer than that of a coniferous forest. Hordes of insects, spiders, snails, and worms live on the forest floor. In early spring, when the new leaves still are not fully grown, large patches of sunlight brighten the forest floor. Wildflowers and ferns grow almost everywhere.

An occasional mouse scurries across your path. Many more small mammals are out of sight under the leaves. A gray squirrel watches you pass and then disappears in the branches. Suddenly, up ahead, you hear a stirring in the underbrush and see a flash of white. A white-tailed deer has spotted you and has dashed away, showing the snowy underside of its tail.

**Figure 3–10** *In the spring and summer, the trees and shrubs in a deciduous forest are green. In autumn, however, the leaves of deciduous plants change color and are then shed.*

**Figure 3–11** *The red fox makes its home in deciduous forests. Salamanders, such as this long-tailed salamander, are also residents of deciduous forests.*

By the side of a rushing stream, you spy a print in the mud. It looks almost like the print of a small human hand. But you know the print was made by a raccoon searching for frogs during the night. Thrushes, woodpeckers, and blue jays flit back and forth between the trees. The ruffled grouse, a relative of the spruce grouse, rests in a tangle of bushes and watches cautiously as you walk by. Under a rotting log, you find a spotted salamander—jet black with big yellow spots. A black snake slithers away as it senses your approach.

By winter, many of the birds have migrated south. (Do you know why?) Snakes and frogs hibernate through the winter. Raccoons, grown fat in autumn, spend the coldest months sleeping in their dens, from which they may emerge during warm spells. The trees in winter are bare, and their branches rattle in the wind. With the coming of spring, however, the leaves will bud and the birds will return. The deciduous forests will come to life once again. But now you are ready to continue your journey.

*Comparing Forest Biomes*

Divide a piece of posterboard into three sections. In each section, create a scene from each of the three forest biomes. You can use crayons to draw the scenes, cut out photographs from magazines, or use any other materials you find helpful. Label each forest scene. Be sure that each forest scene illustrates plants and animals found in that biome.

■ Are any of the plants or animals found in more than one biome?

## Tropical Rain Forests

Your travels now bring you farther south—all the way to the Amazon River of South America. You camp there in a tropical rain forest. Tropical rain forest biomes are also found in central Africa, southern Asia, Hawaii, and even a bit of Australia.

Setting up your tent beneath the dripping trees, you discover that the rain forest is rightly named. In fact, it rains almost every day. Tropical rain forests get at least 200 centimeters of rain yearly. The climate is like summer year round, so plants can grow for all 12 months of the year.

After only a few minutes, your clothes are soaked with dampness and perspiration. The air is muggy and still, although not as hot as you expected. The temperature in the tropical rain forest, or jungle, is hardly ever higher than the temperature on a scorching summer day in Chicago or New York City. Why? The answer is overhead, where the tops of the trees meet to form a green roof, or **canopy,** 35 meters or more above the ground. According to explorer and zoologist Ivan Sanderson, the light below the canopy "is strange, dim, and green." Only along river banks and in places where people or fires have made clearings in the trees does enough sunlight get through the canopy to allow plants to grow on the forest floor.

Most plant life in a rain forest grows in the sunlit canopy. Woody vines called lianas—some thicker than your leg and more than 50 meters long—snake along the branches. You can see orchids and ferns perched on the branches and in the hollows of trees. Tropical rain forests have more varied plant life than any other land biome. The jungle you are exploring has more than 40,000 plant species!

Animal life in the rain forest is also marvelously varied. However, many of the jungle's creatures are out of your sight. High atop the tallest trees, poking here and there above the canopy, sit harpy eagles. Their keen eyes search the canopy below for monkeys and other prey. The canopy is full of parrots, toucans, and hundreds of other colorful birds. At night, bats flit among the trees.

Wild cats called ocelots and 3-meter-long snakes called boa constrictors hunt birds and monkeys in the shorter trees that grow just below the canopy. Standing quietly, you are careful not to startle the tapirs feeding on the ground among the trees. You cannot see the Amazon's big cat, the jaguar, which is very secretive. But as you continue to explore, you think you hear one roaring far off in the jungle. Underfoot, the soil is full of small creatures—centipedes, spiders, ants, and beetles.

**Activity Bank**

Cutting Down the Rain, p.148

**Figure 3–12** *Tropical rain forests are home to more types of living things than all other biomes combined. The large, unusual flowers of a* Heliconia *plant are just one of the strange and beautiful things you might see in a tropical rain forest.*

**Figure 3–13** *The rain forest is home to many animals. The enormous claws of the tamandua help it climb trees and tear open the nests of ants and termites on which it feeds (left). The velvet worm lives among the fallen leaves on the forest floor (right).*

As you make your way back to camp, you hear a strange sound that you cannot identify at first. Then you realize what it is: the ugly noise of a chain saw ripping through the trunk of a giant tree. The most dangerous animal in the jungle is at work! A few seconds later you hear a dull thud as the tree—which has been home to so many birds, monkeys, insects, and frogs—comes crashing to the ground. And you remember that every day an area of tropical rain forest bigger than the city of Chicago is cut down. At that rate, all the rain forests will be gone by the year 2081. Sadly, you prepare to leave the Amazon rain forest, wondering how much of it will be left when you return.

**Figure 3–14** *Tropical rain forests are being destroyed at an alarmingly rapid rate. What happens to the other residents of the forest when the trees are cut down and burned?*

## 3–3 Section Review

1. What are the three forest biomes? How are they different from one another?
2. What are conifers? What are deciduous trees? Give an example of each.
3. What keeps sunlight from reaching the floor of a tropical rain forest?

**Connection—*You and Your World***

4. You probably know that many people in the United States, as well as in other countries, are trying to find ways to save the rain forests. Why do you think the rain forests are being cut down? Why do you think people feel it is important to stop the destruction of the rain forests? Do you think that the rain forests should be saved? Why or why not?

*Saving the Rain Forests*

Organize two teams to debate the following statement: "The destruction of the rain forests will not affect my life."

# 3–4 Grassland Biomes

**Guide for Reading**

*Focus on this question as you read.*

▶ *What is the climate of a grassland biome?*

From the Amazon rain forest, you travel across the Atlantic Ocean until you reach the grasslands of East Africa. **In a grassland biome, between 25 and 75 centimeters of rain fall yearly.** As you might expect from its name, grasses are the main group of plants in a grassland biome. Africa has the largest grasslands in the world, although other large grasslands are found in North America, central Asia, South America, and near the coasts of Australia. Grasslands with a few scattered trees, such as those in Africa, are known as savannas.

Grandeur in the Grass, p.149

Your camp is in a field of grass occasionally dotted with thorny trees called acacias. There are few trees in the grasslands because of the low rainfall. Wildfires, which often rage over the grasslands, also prevent widespread tree growth. And people often set fire to the grasslands on purpose to control the spread of trees.

The animals that roam the grasslands also keep trees from spreading by eating new shoots before they grow too large. As you watch a herd of elephants tearing up acacia trees and feeding on

their leaves, you realize that even large trees are not safe.

Grasses, however, can survive trampling and low rainfall, and still grow thickly. That is why grasslands can feed the vast herds of large herbivores, such as the zebras and antelope you see grazing around you. These animals, in turn, are food for lions, African wild dogs, and cheetahs.

Many mice, rats, and other small animals also inhabit the grasslands, eating seeds, sprouts, and insects. Snakes prowl among the grasses, hunting these creatures. As you walk about your camp in the evening, you take care not to step on a puff adder or other poisonous snake as it searches for prey.

The smaller animals, including snakes, are the prey of the sharp-eyed hawks and eagles that continually sail over the savannas or perch in the acacias. In the distance, vultures circle in the sky, ready to feed on the remains of a zebra killed by lions.

Like the grasslands of North America and many other parts of the world, much of the African savannas has been turned into farms and ranches. And here, as in other parts of the world, overgrazing and overplanting may eventually destroy the grasslands. As you begin the last stage of your journey through the Earth's biomes, you reflect that your next stop—the Sahara Desert—was once a grassland!

**Figure 3–15** *The grasslands of Africa are home to many different kinds of organisms, including acacia trees, giraffes, grasses, antelope, zebras, and ostriches. What other organisms would you expect to see in an African grasslands biome?*

## 3–4 Section Review

1. How much rain does a grassland biome receive each year?
2. Where is the largest grassland biome found?
3. What three factors prevent trees from overrunning grasslands?

**Critical Thinking—*Making Predictions***

4. If all the savannas in Africa are turned into farms and ranches, what effect will this have on the large herbivores that now live on the savannas? How will this affect the carnivores that prey on them?

# 3–5 Desert Biomes

### Guide for Reading

*Focus on these questions as you read.*

- *What is the climate of a desert biome?*
- *What is the difference between a hot desert and a cold desert?*

North of the savannas, the grasslands of Africa become increasingly dry. Eventually you come to the Sahara Desert, which covers almost all of North Africa. In fact, the Sahara is about as big as the entire United States! And it is getting bigger, expanding into the grasslands to the south. **A desert biome is an area that receives less than 25 centimeters of rainfall a year.** Other desert biomes are found in western North America, western Asia, the center of Australia, and along the west coast of South America.

**Figure 3–16** *The Sahara, in northern Africa, is a hot desert (left). Its average annual temperature is about 29.5°C. The Atacama Desert in South America is a cold desert (right). Its average annual temperature is about 17°C.*

## ACTIVITY
### DISCOVERING

*A Desert Terrarium*

Use a wide-mouthed jar to build a small desert terrarium. Add 2 to 5 cm of sand to the bottom of the jar. Plant a few small cacti and other desert plants in the sand. **Note:** *Do not put any animals in your terrarium because it is too small for them.* Cover the mouth of the jar with a piece of screening such as that used in screen doors. Place the jar where it will get plenty of sunlight and heat. Water your desert terrarium no more than once a month.

■ What do you think would happen to the plants in your terrarium if you watered them every day?

Although most people think of a desert as always being hot, a desert can actually be hot or cold. The Sahara is a hot desert—scorching by day, chilly at night. In a cold desert, such as the Gobi Desert in northern China or the Atacama Desert along the west coast of South America, there is also a great difference between daytime and nighttime temperatures. But in a cold desert, daytime temperatures during the winter may be below freezing (0°C)!

Freezing is the last thing you need to worry about as you walk through the burning sands of the Sahara Desert. You cannot help but notice that the plants in the desert are adapted to the lack of rainfall. Many have widespread roots that are close to the surface. This enables the roots to absorb water quickly, before it evaporates. Like the cactus plants of the North American deserts, the aloe plants of the African deserts have thick, fleshy stems that help them store water. After a rainfall, the stem of an aloe plant swells to almost 3 meters in diameter. You may be familiar with aloe as an ingredient in many soaps and hand lotions. Aloe is extremely useful in soothing skin irritations and minor burns (including sunburn).

Even though you look hard, you see few animals in the Sahara. But they are there. By day, lizards and small rodents often escape the heat in underground burrows. Here the temperature may be as much as

**Figure 3–17** *The orange and yellow flowers of an aloe plant (left) and the grass and trees of an oasis (right) contrast sharply with the bleak desert that surrounds them.*

30°C cooler than at the surface. Night brings the animals to the surface searching for food.

Like the plants, desert animals must live on as little water as possible. Most of the water used by desert animals comes from the seeds and stems of plants, which are about 50 percent water. The most famous desert animals—camels—can live without water if there are enough plants available for them to eat. In fact, camels can get along without water for up to 10 days! During this time, they live off the water stored in the body fat in their humps. And like other desert animals, camels lose almost no water in their wastes. Only in these ways can they survive in a world where rain hardly ever falls.

As your trek across the dry Sahara comes to an end, you realize you have completed your journey through the Earth's major land biomes. That means you are now ready to explore the largest biome on Earth—the oceans.

**Figure 3–18** *Certain animals have evolved in ways that help them to survive in the harsh conditions of desert biomes. The chameleon holds its body as far from the ground as possible as it tiptoes along (top left). The sidewinding movement of the adder is also an adaptation that minimizes contact with the hot sand (top right). What are some of the adaptations camels have for surviving in the desert (bottom)?*

## 3–5 Section Review

1. Describe the main characteristics of a desert biome.
2. What is the difference between a hot desert and a cold desert?
3. In what ways are desert plants and animals adapted to the lack of rainfall? Give two examples of each.

**Critical Thinking—*Making Inferences***

4. Why do you think there are no tall trees in the desert?

**Guide for Reading**

*Focus on this question as you read.*

- *How are the Earth's two major water biomes classified?*

# 3–6 Water Biomes

Your trip around the world would not be complete without a visit to the water biomes. After all, most of the Earth's surface is covered with water. **The two major water biomes are the marine biome and the freshwater biome.**

**Figure 3–19** *Starfish, flowerlike sea anemones, and seaweed are some of the marine organisms that live at the edge of the sea.*

## The Marine Biome

The **marine biome,** or ocean biome, covers about 70 percent of the Earth. Organisms that live in this biome have adaptations that allow them to survive in salt water. Other factors that affect ocean organisms are sunlight, temperature, water pressure, and water movement. The oceans can be divided into different zones, or areas, based on these factors. Each of these zones contains organisms that are adapted to conditions in that zone.

You begin your exploration of the ocean near the shore. Most marine organisms live near the surface or near the shore. Animals that live near the shore are alternately covered and uncovered by the tides. Many of the animals burrow into the sand while others attach themselves to rocks to keep from being washed out to sea. Strolling along the shore you find clams, barnacles, and sea stars in shallow tidepools.

Past the low-tide line, algae and microscopic plants called **phytoplankton** live near the surface of the ocean where they can receive the most sunlight. They use the sunlight to produce food. Almost all the animals in the ocean depend either directly or

**Figure 3–20** *The enormous seaweed known as kelp forms vast marine "forests," which are often home to the playful sea otter.*

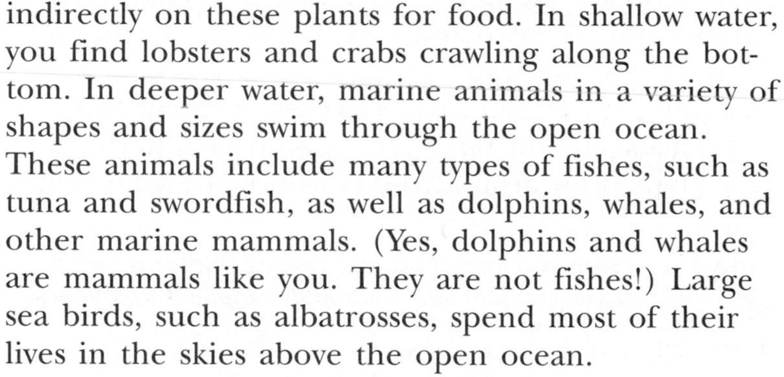

indirectly on these plants for food. In shallow water, you find lobsters and crabs crawling along the bottom. In deeper water, marine animals in a variety of shapes and sizes swim through the open ocean. These animals include many types of fishes, such as tuna and swordfish, as well as dolphins, whales, and other marine mammals. (Yes, dolphins and whales are mammals like you. They are not fishes!) Large sea birds, such as albatrosses, spend most of their lives in the skies above the open ocean.

At one time, scientists thought that the deepest parts of the ocean had no life at all. The deep ocean is an area of cold temperatures, high pressures, and complete darkness. Now, however, scientists have discovered that some of the strangest marine organisms live in the deep ocean. Many of them have unusual adaptations for survival in this dark environment. For example, some deep-sea fishes and squid actually have organs that are capable of producing light. In 1974, scientists discovered previously unknown organisms—giant tube worms, blind crabs, and huge clams—clustered around hot-water vents on the ocean floor. These strange creatures do not rely on energy from the sun to survive. Instead, they use chemicals from deep inside the Earth.

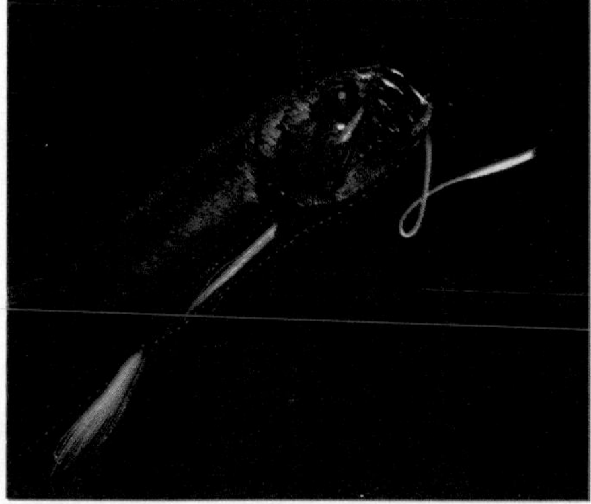

**Figure 3–21** *Many marine animals are creatures of the open sea. The albatross spends most of its life soaring above the waves (center). The sailfish may use its long bill to kill or injure prey (top right). Dolphins are not fish, but mammals (left). Deep-sea fishes are monstrous in appearance, but not really in size (bottom).*

## The Freshwater Biome

From the depths of the oceans, you move on to the Earth's other water biome. The **freshwater biome** includes both still water and running water. Lakes and ponds are still water. Streams and rivers are running water.

As in the marine biome, there are a number of factors that affect freshwater life. These factors are temperature, sunlight, the amount of oxygen and food available, and the speed at which the water moves. These factors determine which organisms live in a freshwater environment.

**Figure 3–22** *The young mayfly lives in fast-moving freshwater streams. Although young mayflies live in aquatic habitats, adult mayflies are winged creatures of the air.*

Walking along a fast-moving stream, you find that the organisms in the stream have special structures that keep them from being swept away. Many plants have strong roots that anchor them to the stream bottom. Others have stems that bend easily with the moving water. Mosses cling to rocks in the stream. And the young of some insects have suckerlike structures on their bodies that help them to attach themselves to rocks or other objects in the stream. Fishes such as trout have streamlined bodies for swimming in the fast-moving water.

Leaving the stream, you next visit a small lake. Here such common freshwater plants as waterlilies and cattails grow around the shore of the lake, while

**Figure 3–23** *Frogs (bottom right) and grebes (top) live in still freshwater habitats (bottom left). Although excellent swimmers, grebes can barely walk on land. How does building a floating nest of reeds, such as the one shown here, help grebes survive?*

algae and duckweed float on the surface. In the middle of the lake, a fish—perhaps a yellow perch or a bluegill—breaks the surface. A water snake glides silently by. You notice the bulging eyes of a frog staring out at you from among the duckweed. You cannot see the microscopic plants and animals that are a part of any freshwater biome. You do catch a glimpse of a family of ducks and a shy raccoon, however. These animals visit freshwater biomes to feed or nest. What other animals might you see in a freshwater biome?

*Life Near a Trout Stream*

Many writers have been influenced by nature. Sean O'Faolain (1900–1991) was born in Ireland and often wrote stories about life in the Irish countryside. You might enjoy reading his short story "The Trout."

## Estuaries

After leaving the lake, you move on to the last stop on your journey, the Chesapeake Bay on the eastern coast of the United States. Now you are at an **estuary** (EHS-tyoo-air-ee)—the boundary between a freshwater biome and a marine biome. The Chesapeake Bay is the largest estuary in the United States. Estuaries include salt marshes, lagoons, mangrove swamps, and the mouths of rivers that empty into an ocean. Estuaries are areas that contain a mixture of fresh water and salt water. Some scientists think that estuaries make up a separate biome, while others consider them as ecosystems.

Because estuaries are usually shallow, sunlight can reach all levels of the water. Marsh grasses, algae, and other kinds of plants live in estuaries and provide food for a variety of fishes, crabs, oysters, and

**Figure 3–24** *Estuaries include salt marshes (left) and mangrove swamps (right). The spreading roots of the mangrove tree help to keep the plant from falling over.*

shrimp. Estuaries are especially important as "nurseries" for many different types of young fishes and other animals before they head out to the open ocean. Many sea birds also nest in estuaries.

An estuary such as the Chesapeake Bay is very fertile. Because the Bay produces so many crabs, oysters, and fishes, it is an important part of the economy in Virginia and Maryland. Many people also find the Bay a pleasant area in which to live or spend their leisure time.

As you look back on your trip through the Earth's biomes, you wonder what will become of estuaries—and of forests, deserts, and grasslands—in the future. Can people enjoy the Earth's wild places without destroying them? What can you do to help protect the Earth's biomes?

**Figure 3–25** *Chesapeake Bay, like other estuaries, is an extremely fertile habitat. Blue crabs and other kinds of seafood are abundant in the Bay. Why is important to protect the Bay and other estuaries from pollution and overuse?*

## 3–6 Section Review

1. Describe the two major water biomes.
2. Compare the factors that affect marine life with the factors that affect freshwater life.
3. Why are there no green plants on the bottom of the ocean?
4. Why are estuaries important?

**Connection—*You and Your World***

5. What effect might population growth in the area around the Chesapeake Bay have on the ecology of the Bay? How might this affect the area's economy? What do you think can be done to protect the Bay?

# CONNECTIONS

## *Fish Farming*

What do you think of when you hear the word farm? You may think of neat rows of corn, chickens scratching in a barnyard, or cows grazing in a meadow. But do you ever think of fishes?

Fish farming, or aquaculture, has become a booming business in the United States. Americans are increasingly concerned about their *health* and are eating more fish and less meat. Today, about one fourth of all the fish and other kinds of seafood served in restaurants or sold in supermarkets has been farmed rather than caught in a lake or stream.

The most popular fish being farmed in the United States is catfish. Part of the appeal of catfish is the popularity of Cajun food from Louisiana—blackened catfish is a favorite dish in this culture. The second most popular fish is trout, which was the first fish to be widely farmed in the United States. Trout farmers have combined aquaculture with new breeding techniques to create "efficient" fishes that mature in 10 months rather than in the natural 18. Trout farmers hope to soon breed fishes that mature even faster.

Fish farming is done both outdoors and indoors. Outdoor farming may be done in special ponds or in portions of the ocean that have been sectioned into pens. Indoor fish farming takes place in large tanks that look like giant aquariums. Indoor fish farmers are even able to raise tilapia, a popular tropical fish that is native to the Nile River in Egypt.

Other kinds of seafood farmed in the United States include shrimp and salmon. Salmon are farmed mainly in Washington State and Maine, where they are raised in floating pens in saltwater bays. Shrimp are farmed mostly in Texas, in ponds in the Rio Grande Valley. Fish farming may become even more widespread as more people realize that fish is a healthful and tasty alternative to red meat.

# Laboratory Investigation

## Building a Biome

### Problem

How do different plants grow in different biomes?

### Materials *(per group)*

| | |
|---|---|
| 2-L cardboard milk carton | scissors |
| sandy soil or potting soil | lamp |
| 5 lima bean seeds | index card |
| 30 rye grass seeds | tape |
| 10 impatiens seeds | stapler |
| clear plastic wrap | |

### Procedure

1. Your teacher will assign your group one of the following biomes: desert, grassland, deciduous forest, rain forest.
2. Cut away one side of a milk carton. Poke a few small holes in the opposite side for drainage. Staple the spout closed.
3. Fill the carton with soil to within 3 cm of the top. **Note:** *If your group has been assigned the desert biome, use sandy soil.*
4. At one end of the carton, plant impatiens seeds. In the middle of the carton, plant lima bean seeds. Scatter rye grass seeds on the soil at the other end of the carton.
5. On an index card, identify your group, the seeds planted, and the type of biome. Tape the card to the carton.
6. Water the seeds well. Cover the open part of the carton with plastic wrap.
7. Put the carton in a warm place where it will remain undisturbed. Observe daily.
8. After the seeds have sprouted, follow the instructions for your group's biome:
   Desert: Let the soil dry to a depth of 2.5 cm; 5 to 6 hours of light per day.
   Grassland: Let the surface dry, then add water; 5 to 6 hours of light per day.
   Deciduous forest: Let the surface dry, then add water; 1 to 2 hours of light per day.
   Rain forest: Keep the soil surface wet; no direct light.
9. Observe the development of the plants in the biomes of all the groups.

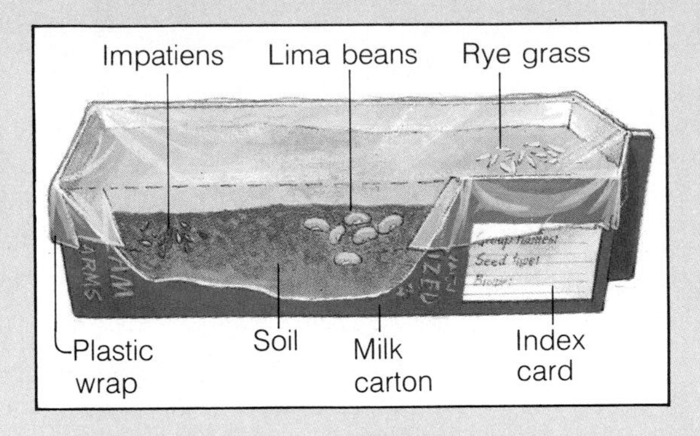


### Observations

1. After the seeds have grown for a week, describe the growth in each biome.
2. In which biome did most of the seeds grow best?
3. Where did the rye grass seeds grow best? The lima beans? The impatiens?
4. Which plants grew well in more than one biome?
5. How do lima beans react to little light?

### Analysis and Conclusions

1. Explain why the plants grew differently in each biome.
2. Why did the seeds need water when they were planted?
3. What was the variable in this experiment?
4. **On Your Own** Predict how the impatiens, lima bean, and rye grass seeds would grow in tundra and coniferous forest biomes. Design an experiment to test your prediction.

## Summarizing Key Concepts

### 3–1 Biogeography

▲ Biogeography is the study of where plants and animals live throughout the world.

▲ Biomes are divisions based on similar climate, plants, and animals.

▲ The major land biomes are tundras, coniferous forests, deciduous forests, tropical rain forests, grasslands, and deserts.

### 3–2 Tundra Biomes

▲ Tundra biomes are very cold and dry.

▲ Most of the water on the tundra is permanently frozen in the soil as permafrost.

### 3–3 Forest Biomes

▲ The three major forest biomes are coniferous forests, deciduous forests, and tropical rain forests.

▲ Trees in a coniferous forest are conifers, which produce seeds in cones.

▲ Deciduous trees shed their leaves in the autumn and grow new leaves in the spring.

▲ Tropical rain forests have more varieties of plants and animals than any other land biome.

### 3–4 Grassland Biomes

▲ Grassland biomes receive between 25 and 75 centimeters of rain yearly.

▲ Low rainfall, fires, and grazing animals prevent the widespread growth of trees on grasslands.

### 3–5 Desert Biomes

▲ Deserts receive less than 25 centimeters of rain yearly.

▲ Deserts can be either hot or cold.

▲ Plants and animals in a desert are adapted to the lack of rainfall.

### 3–6 Water Biomes

▲ The two major water biomes are the marine biome and the freshwater biome.

▲ The marine, or ocean, biome covers about 70 percent of the Earth.

▲ The freshwater biome includes both still water (lakes and ponds) and running water (streams and rivers).

▲ An estuary is an area that contains a mixture of fresh water and salt water.

## Reviewing Key Terms

*Define each term in a complete sentence.*

### 3–1 Biogeography

biogeography
dispersal
biome

### 3–2 Tundra Biomes

permafrost

### 3–3 Forest Biomes

conifer
taiga
canopy

### 3–6 Water Biomes

marine biome
phytoplankton
freshwater biome
estuary

# Chapter Review

## Content Review

### Multiple Choice

*Choose the letter of the answer that best completes each statement.*

1. The freshwater biome includes all of the following except
   a. lakes. b. streams.
   c. ponds. d. oceans.
2. The forest biome that reaches farthest north is the
   a. rain forest. b. coniferous forest.
   c. deciduous forest. d. savanna.
3. You would probably expect to find caribou living in a
   a. rain forest biome.
   b. grassland biome.
   c. desert biome.
   d. tundra biome.
4. All of the following are examples of estuaries except
   a. mangrove swamps.
   b. rivers.
   c. salt marshes.
   d. lagoons.
5. Anything that prevents plants and animals from moving from place to place is called a
   a. biome. b. habitat.
   c. barrier. d. community.
6. Most living things in the marine biome are found near the shore or
   a. in fast-moving streams.
   b. in the deep ocean.
   c. near the ocean surface.
   d. in small ponds.
7. Not many trees grow in a grassland biome because of low rainfall, fires, and
   a. animals.
   b. freezing temperatures.
   c. high winds.
   d. floods.
8. The greatest variety of plant and animal species is found in a(an)
   a. desert. b. estuary.
   c. rain forest. d. taiga.

### True or False

*If the statement is true, write "true." If it is false, change the underlined word or words to make the statement true.*

1. Horses dispersed from North America into Asia by crossing a <u>mountain range</u> between Alaska and Siberia.
2. Lakes and ponds are part of the <u>marine</u> biome.
3. A climate that is too cold for an organism to survive is an example of an <u>ecological barrier</u>.
4. Trees that shed their leaves in autumn are called <u>conifers</u>.
5. Most of the plant life in a tropical rain forest can be found growing <u>on the forest floor</u>.

### Concept Mapping

*Complete the following concept map for Section 3–1. Refer to pages G6–G7 to construct a concept map for the entire chapter.*

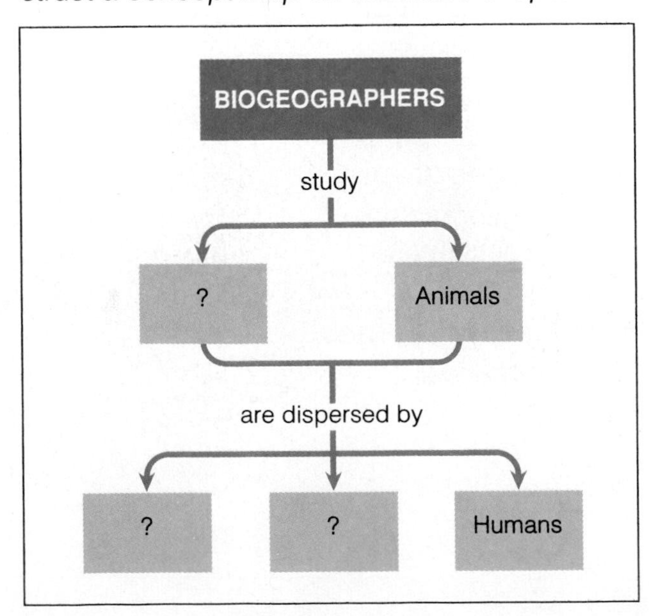

## Concept Mastery

*Discuss each of the following in a brief paragraph.*

1. How do barriers prevent plants and animals from spreading into new areas? Give three examples.
2. Why is climate an important factor in dividing the Earth into biomes?
3. Why do you think it is difficult for scientists to agree on the number and kinds of biomes?
4. Briefly describe each of the six major land biomes.
5. Why are most organisms that live in a marine biome unable to survive in a freshwater biome?
6. What kinds of adaptations are needed by organisms living in the deepest parts of the ocean?

## Critical Thinking and Problem Solving

*Use the skills you have developed in this chapter to answer each of the following.*

1. **Making generalizations** Why is it difficult to tell exactly where one biome ends and another begins?
2. **Relating cause and effect** In which biome would you expect to find the animals shown below? What effect do they have on their environment?

3. **Relating facts** A tropical rain forest has a greater variety of species than any other biome. What characteristics of a tropical rain forest could account for this?
4. **Making predictions** What do you think would happen to the lions and other carnivores on the African savannas if a disease killed all the herbivores, such as zebras and wildebeests?
5. **Making inferences** What characteristics would you expect animals that remain on the tundra all year instead of moving south for the winter to have?
6. **Relating concepts** An estuary is an area where fresh water and salt water meet. Explain why it is difficult for scientists to classify estuaries as either freshwater or marine biomes.
7. **Using the writing process** Imagine that you are a reporter for your local newspaper. You have been assigned to interview a rancher in South America who wants to clear an area of tropical rain forest in order to provide grazing land for cattle. Write out a list of questions that you will ask the rancher in your interview.

# Wildlife Conservation

## Guide for Reading

*After you read the following sections, you will be able to*

**4–1 Identifying Problems**

- Discuss the reasons for the extinction of organisms.
- Explain why people should try to save endangered species.

**4–2 Seeking Solutions**

- Describe some conservation measures aimed at saving wildlife.

As the noisy truck approaches, the rhinoceros begins to run away. But it is not fast enough. One of the people in the truck picks up a rifle, aims, and fires. The rhino stumbles and falls to the dusty ground of the African plain. The people drive up to the fallen rhino. One of them takes out a saw and begins to cut off the rhino's horns.

Rhino horns are nothing more than large curved cones made of the same substance as your fingernails. But in some parts of the world, people believe that rhino horns have magical properties. So the horns are worth more than their weight in gold.

Fortunately for this one rhinoceros, things are not always what they seem. The rhino was shot by a tranquilizer dart and is fast asleep. Its horns are being removed by skilled game wardens. Without its horns, the rhino looks rather strange—but it is now safe from illegal hunters. It is also still safe from other predators, because it is the rhino's size and not its horns that deters attacks.

Removing rhino horns is one of the more unusual ways of protecting wildlife. What are some other things that people do to help save rare organisms? Why do organisms become rare in the first place? Read on to find out the answers to these questions.

### Journal *Activity*

***You and Your World*** Explore your thoughts and feelings about dinosaurs and other organisms that have vanished from the Earth.

*The black rhinoceros peers nearsightedly at an uncertain future. In the past 20 years, over 90 percent of Africa's black rhinoceroses have been killed for their horns.*

**Guide for Reading**

*Focus on these questions as you read.*

- ▶ *What is extinction? How do organisms become endangered or extinct?*
- ▶ *Why should people care about endangered species?*

**Figure 4–1** *A 1990 reconstruction shows the dodo as a sleek bird with a small, dignified tail. This is quite a change from the obese, stupid bird with a silly feather-duster tail in John Tenniel's illustration for Lewis Carroll's* Alice in Wonderland.

# 4–1 Identifying Problems

Two thousand kilometers east of Africa, in the warm tropical waters of the Indian Ocean, lies the small island of Mauritius. This island is best known for the dodo—a large, fat, flightless bird that once lived there. You may know about the dodo from jokes or expressions such as "dumb as a dodo." But the real story of the dodo is not funny at all.

After Mauritius was discovered by Europeans in the early sixteenth century, ships began stopping there regularly to pick up fresh supplies of food and water. Huge numbers of dodos were killed by sailors for food and for sport. (A twentieth-century person might not think it sporting to walk up to a practically tame bird and hit it over the head!) Amazingly, the dodo was able to survive a hundred years of assault by sailors armed with clubs. But it was not able to survive thirty years of the first permanent human settlement on Mauritius.

The people who settled on Mauritius brought with them dogs, pigs, cats, monkeys, and rats. (The settlers probably did not intend to bring rats, but rats manage to accompany humans almost everywhere.) The pigs, cats, monkeys, and rats killed and ate the dodos' eggs and chicks. The dogs killed the adult birds. The settlers themselves killed all the dodos they could find, even though the birds were quite harmless and not worth eating. By 1680, dodos had become **extinct.** In other words, their species no longer existed.

**The process by which a species passes out of existence is known as extinction.** Extinction is a natural part of our planet's history. In fact, more than 99 percent of all the living things that have ever existed on Earth are now extinct. Since life began on Earth more than 3.5 billion years ago, countless species of microorganisms (microscopic organisms), fungi, plants, and animals have appeared, survived for a time, and then passed out of existence. For the last 600 million years, species have become extinct at the average rate of about one per year. But in the past three hundred years or so, human activities—such as hunting, farming, building cities, and cutting down forests—have greatly increased the rate of extinction.

Experts estimate that the extinction rate is now several species per day. And unless current trends are stopped, the extinction rate could be up to several species per hour by the end of the century!

Human activities increase the rate of extinction because they change the environment too quickly for organisms to adapt. Adaptation occurs through the slow process of evolution. And in evolutionary terms, something that takes place over a period of hundreds—even thousands—of years occurs quickly. (Although humans are not responsible for all the quick changes that occur, natural events that cause sudden changes are few and far between.)

Let's return for a moment to the example of the dodo. Like organisms the world over, the dodo evolved in response to the challenges of its environment. These challenges did not include predators such as humans, dogs, cats, pigs, monkeys, and rats. When humans settled on Mauritius, they made a major change in the island's environment—in the blink of an evolutionary eye, humans added fierce predators to the island's ecosystems. Unable to change quickly enough to survive the challenge of the predators, the dodo became extinct. So did giant tortoises, owls, and many other native organisms.

When their environment changes for the worse, organisms become rarer and rarer. Organisms that are so rare that they are in danger of becoming extinct are said to be **endangered.** About 4500 kinds of animals and 20,000 kinds of plants are endangered. In addition, hundreds of species are not known well enough to be formally classified as endangered.

*A Picture Says a Thousand Words*

Create a poster to help make other students aware of the problems of wildlife extinction.

**Figure 4–2** *Dinosaurs, such as* Triceratops, *became extinct millions of years ago. Bones that have turned to stone are all that remain (right). An artist's reconstruction shows what* Triceratops *might have looked like when it was alive (left).*

**Figure 4–3** *The Texas blind salamander (top right), sifaka (bottom right), aye-aye (top left), and Knowlton cactus (bottom left) are all endangered species. The aye-aye is in particular trouble because people consider it to be bad luck and kill it on sight. What does it mean to say a species is endangered?*

The human activities that cause species to become endangered or extinct are many. They also are the results of a variety of motives. In this textbook, we have chosen to divide the human activities that threaten wildlife into two broad categories based on their main purpose. The first category includes human activities whose main purpose is to kill specific kinds of wildlife. We have called this category "intentional killing." The second category includes activities that are not specifically directed at killing wildlife. These activities may kill as many or more living things as purposeful killing, but their main purpose is not the death of organisms. Such activities include cutting down forests, introducing foreign species, and running over animals in boat or car accidents.

## Intentional Killing

The rhinoceros you read about in the chapter opener belongs to a species that numbered about 65,000 in 1970 and about 3000 in 1991. Almost all of the rhinos that died during that 21-year period were killed by poachers (illegal hunters). A poacher is able to sell a rhino horn for about $200. This may not seem like a lot of money. But in the poorest countries of Africa, the average person makes only about $200 a year—which may not be enough to obtain necessities such as food and shelter. Can you see why it is not always easy to stop overhunting of wild animals?

Rhinos have been overhunted for their horns, which are used to make dagger handles in Yemen and folk remedies in Asia. Other animals have been overhunted for other reasons. Some—such as wolves and bald eagles—were shot, poisoned, and trapped because they were believed to prey on humans and livestock. Others—sun bears, fruit bats, whales, and sea turtles, to name a few—are killed for gourmet food. Still others are killed to fulfill the demands of fashion. Elephants are shot for their tusks, which are used to make ivory jewelry and trinkets. The endangered hawksbill turtle is hunted for its shell, which is used to make jewelry, and for its meat. Snow leopards, sea otters, and wild chinchillas are among the many animals that became endangered because their beautiful skins were in great demand for coats, hats, and other fur products. A fad for shoes, handbags, and other leather goods made from the tough, bumpy skins of American alligators came very close to killing off the species. Snowy egrets were nearly hunted into extinction for their lovely feathers, which were used to decorate hats.

**Figure 4–4** *The sun bear gets its common and scientific names from the sun-colored crescent on its chest. It is endangered due to habitat destruction, the fur and pet trades, and its grilled paws being regarded as a delicacy. The snow leopard has been hunted to the brink of extinction for its fur. The passenger pigeon was once the most numerous bird. But during the 1800s, it was slaughtered by the millions. The last passenger pigeon died in a zoo in 1914.*

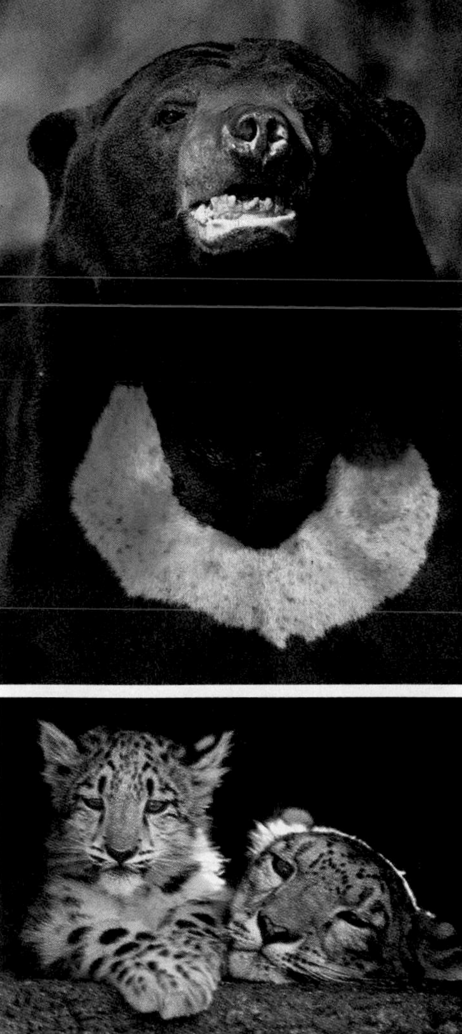

Figure 4–5 *The beautiful golden-brown wood of the Hawaiian koa tree was once used extensively. Now the vast koa forests are gone.*

Up to this point, you have been reading about ways in which uncontrolled killing can lead to animal species becoming endangered. Uncontrolled killing can also cause the downfall of plant species. In Australia, for example, there were once huge forests of red cedar trees. The largest red cedar trees were cut down for their wood, which was used to make furniture. Soon there were no trees large enough to harvest for their wood. But the destruction did not stop there. Because cattle cannot eat red cedar trees, ranchers made a special effort to destroy all the red cedar trees they could find as they cleared away forests to make room for pastures. Now only a few trees remain.

## Destroying Habitats

The examples you just read about involve living things that became endangered because they were (or are) being killed on purpose. However, far more species are in trouble because of the destruction of their habitats. (Recall from Chapter 1 that a habitat is the place where an organism lives and obtains the resources it needs to survive.) As you now read about habitat destruction, keep in mind that these are but a few of the many examples. Habitats have been damaged or destroyed, and continue to be damaged or destroyed, in every one of the biomes you learned about in Chapter 3.

Figure 4–6 *This tropical forest was drowned by the construction of a dam in Brazil. Although quite useful to humans, dams destroy habitats up river, down river, and within the river itself.*

**DEFORESTATION** People cut down forests to obtain wood or to clear land for farms, factories, shopping malls, office buildings, and homes. The removal of forests is known as **deforestation.** History shows that deforestation has been going on for a long time and has occurred in most of the countries of the world.

When Europeans first arrived in the New World, forests covered most of the eastern half of the United States. Almost all the original forest was cleared to make room for farms and towns and to harvest timber. When the old forests were destroyed, the plants and animals that had evolved to live in those environments had no place to go. Some, such as the ivory-billed woodpecker and eastern bison, became extinct. Others, such as the Oconee bells shown in

Figure 4–8 on page 106, are found in only a few places and are rare or endangered.

Today, forests in the western part of the United States are being cut. Many people are concerned that some of the rare species of the Pacific Northwest and Alaska—such as the spotted owl, Pacific yew, and American marten—will become extinct as their habitat shrinks.

The deforestation of greatest concern to people around the world is occurring in the tropical rain forests of Latin America, Africa, and Asia. Each year, an area of tropical rain forest the size of Washington state disappears. About 40 percent of the Earth's tropical rain forests are already gone, and the rate of deforestation is increasing. Some tropical forest is destroyed as hardwood trees—such as teak and mahogany—are harvested. Much more is cut and burned to make room for farms and cattle pastures. The newly cleared land is productive at first. But most of the land stops producing enough food or grass within four to eight years. New areas of forest must be cut in order to feed people and livestock. And as the forest shrinks, the plants, animals, and other organisms that live in, on, or among the trees disappear.

**Figure 4–7** *Large areas of forest in the western United States—including Washington (top left) and Alaska (bottom left)—are being cut down. But the deforestation of greatest concern to most people is taking place in the tropics (right). Why are people more concerned about tropical forests?*

**Figure 4–8** *Deforestation affects more than trees. What will happen to the two-toed sloth (top right), margay (bottom right), toad and black orchid (top left), cottontop tamarin (center left), and Oconee bells (bottom left) if their forest homes are destroyed?*

In Chapter 1, you learned that all ecosystems are interconnected. As you might expect, deforestation causes a great deal of damage to nearby ecosystems. For example, the burning of one area of forest in West Africa causes acid rain that damages other areas of the forest. Deforestation also hurts more distant ecosystems. When hills are stripped of their covering of trees, dirt that is usually held in place by plant roots can be washed into lakes and rivers, damaging freshwater biomes. Eventually, muddy rivers carry their load of dirt to the ocean. The excess dirt can then harm marine biomes. Ultimately, deforestation affects ecosystems all over the world. For example, when forests are cleared by burning, carbon dioxide is released into the air. This increases the amount of carbon dioxide in the air. Can you predict how deforestation affects the carbon and oxygen cycles all over the Earth?

**DESERTIFICATION** What do you think of when you hear the word desert? You may think of cactus plants and roadrunners. Or you may think of camels and shifting sand dunes. You probably do not think of grassy fields and grazing herds of goats, sheep, and cattle. Yet when too many animals graze in an area, grassland may be transformed into a desert. The process in which desertlike conditions are created where there had been none in the recent past is known as **desertification** (dih-zert-uh-fih-KAY-shuhn).

A little desertification occurs naturally at the places where deserts meet other biomes. If rainfall is plentiful for several years, the desert may shrink a bit; if rainfall is scarce, the desert may expand a bit. In recent years, however, more and more desertification has occurred as the result of human actions such as growing crops, raising livestock, and cutting down forests.

Unlike the natural deserts that you learned about in Chapter 3, deserts made by human actions are barren and lifeless. Sometimes, desert organisms move into a new desert and make it their home. Once in a while, the area may gradually return to its former state if there is enough rain and if the grazing animals are kept away. Too often, however, the newly made desert remains an empty wasteland.

*The Green Literary Scene*

"Knowledge is power," so the saying goes. A good knowledge of environmental issues will help to give you the power to change things for the better. Here are a few books to help you get started: *Going Green: A Kid's Handbook to Saving the Planet*, by John Elkington, Julia Hailes, Douglas Hill, and Joel Makower; *A Kid's Guide to How to Save the Planet*, by Billy Goodman; and *The Population Explosion*, by Paul Ehrlich and Anne Ehrlich.

**Figure 4–9** *Overgrazing is one of the factors that transformed green pastures into barren desert in northeast Africa. What is the process of making a desert called?*

**Figure 4–10** *Wetlands, such as this marsh in the Louisiana bayous, are home to many creatures. The survival of organisms such as the roseate spoonbill (left), whooping crane (center), and Everglades kite (right) is made uncertain by continued wetlands destruction.*

**WETLANDS DESTRUCTION** Wetlands are exactly what their name suggests—wet lands, such as swamps, marshes, and bogs. At one time, wetlands were considered nothing more than ugly breeding grounds for mosquitoes. Now people realize that wetlands are extremely valuable ecosystems. They are temporary homes for migrating waterbirds, and permanent homes for minks, alligators, mangroves, Venus' flytraps, frogs, turtles, and many other organisms. Wetlands are also the source of nutrients for many ocean biomes near the shore.

Unfortunately, about half the Earth's wetlands have been drained, filled in, or destroyed by pollution. The effects of wetlands destruction are far reaching. For example, farmers in southern Florida are draining marshes to grow more crops. The marshes contain snails eaten by small hawklike birds known as Everglades kites. As the marshes vanish, so does the kites' food supply. The rest of the story should be familiar to you by now. Yes, the Everglades kite is close to extinction. What are some other possible effects of continued wetlands destruction on the organisms that live there? How might this destruction affect fishing and duck hunting?

**POLLUTION** As you can see in Figure 4–11, pollution can be a threat to living things. Pollution comes in all shapes and sizes. Birds and useful insects may be poisoned when crops are sprayed with chemicals meant to kill pests. Acid rain can kill water plants, deform fish, and prevent fish eggs from developing. Birds, sea turtles, fishes, and other animals can become hopelessly tangled in bits of discarded plastic fishing lines and fishing nets. Can you think of other ways in which pollution harms wildlife?

**Figure 4–11** *Because synthetic substances such as plastics break down extremely slowly, they can be dangerous to wildlife for a long time. This seal tangled in nylon net survived its ordeal. Most entangled organisms are not so lucky.*

## Changing Communities

The sad story of the dodo illustrates what can happen when foreign species are introduced to an organism's environment. Species that are released into a place where they had not previously existed are known as **exotic species.** Don't be confused by this use of the word exotic—exotic species may not be strangely beautiful or different in a way that makes them striking or fascinating. Many exotic species are quite ordinary—the pigeon that you might see pecking at crumbs in a city park is an exotic species.

Exotic species may directly interact with the native species. This interaction may take the form of competition or predation—some examples of which you shall now read about.

If you are from the South, you are probably familiar with the kudzu vine. This vine, imported from Japan to feed sheep and goats, grows extremely rapidly all over everything in sight, including other plants. Kudzu competes with native plants for light, water, and nutrients. Eventually, the kudzu vine causes the death of the plants it has grown over by preventing them from getting enough light.

Exotic predators have been the downfall of many species, including the dodo. Pigs, rats, mongooses, cats, and dogs have each at some time been the main cause of extinction of at least two species of birds. Herbivores, which can be thought of as predators of plants, can also bring about the extinction of their prey. In Hawaii, goats and cattle have eaten several types of plants into extinction.

*Once, a Once-ler . . .*

What happens when careless greed has its way?
Why couldn't the Bar-ba-loots stay?
Why did the Swomee-Swans fly away?
How does the Once-ler cook his own goose?
Read *The Lorax*, by Dr. Seuss.

**Figure 4–12** *Exotic species can be the downfall of organisms. Kudzu vines have completely covered these trees in North Carolina (top left). The brown tree snake has killed off most of the birds of Guam (top right). The Indian mongoose, brought to Hawaii and the Caribbean islands to kill rats, found that native birds were more to its taste (bottom).*

Have you ever arranged dominoes in a long line and then tapped the first one? If so, you might have observed that as the first domino fell, it knocked over the second domino, which knocked over the third, and so on. In a similar way, when one species is killed, other species may be brought down with it. By interfering with the normal interactions in a community, exotic species can act like a finger tapping the first domino in a line. Thus exotic species can cause native organisms to become endangered or extinct even if the exotic species do not interact directly with them. When the goats and cattle ate the Hawaiian plants into extinction, they indirectly caused the extinction of certain native birds that fed on the plants. The extinction of the birds in turn caused the extinction of other kinds of plants, which depended on the birds to pollinate them.

## Accidental Killing

Imagine an enormous gray animal with the round, fat body and front flippers of a porpoise, a paddlelike tail, beady eyes, and a rather "cute" face. This strange-looking creature is the manatee, an endangered aquatic animal that lives in rivers in Florida. It is a peaceful, slow-moving animal that spends its time near the surface of the water, grazing on floating plants. Unfortunately for the manatee, people have discovered that its habitat is ideal for whizzing around in motorboats. A collision with a motorboat's sharp, whirling propeller can badly injure or even kill a manatee. In the first six months

**Figure 4–13** *Government agents confiscated these parrots before the smuggler had a chance to sell the survivors. Trained whales and dolphins have helped to make people concerned about wild whales and dolphins. However, this benefit is not without its price. Most dolphins do not survive more than two years in captivity.*

of 1990, more than 10 percent of Florida's manatees died. About one third of these deaths are known to have been caused by collisions with motorboats.

Very young children sometimes have to be warned not to hug the family pet too tightly, least they "love it to death." Strange as it may seem, some apparently harmless activities meant to increase people's appreciation for nature may pose a threat to the Earth's living things. People are loving wildlife to death!

Many people enjoy growing strange and beautiful plants in their gardens and homes. But some plants—certain kinds of cactuses, orchids, and tulips, for example—are collected from the wild. Overcollection has made a number of species rare or endangered in their natural habitat.

Owning unusual pets—such as parrots, monkeys, and saltwater fish—may help people feel closer to nature. But potential owners should be aware that some "pet" animals are taken from the wild. This unwise and selfish practice can also be extremely cruel and wasteful. For example, baby parrots can be captured by cutting down the trees that contain their nests. As you can imagine, very few baby birds survive the crash to the forest floor. For every 100 parrots taken from the wild, only about 10 survive long enough to be sold.

**Figure 4–14** *Recently, two new laws were put into effect. One limits the number of tourists who can visit the Galapagos Islands each year (bottom). The other prohibits whale-watching tours from getting close to the whales (top). Why were these laws enacted?*

## What Does This Mean to Me?

Figure 4–15 *Some modern medicines come from unusual wildlife sources, including snake venom and the rosy periwinkle plant.*

You have just learned about the many ways in which wildlife is threatened by human activities. And perhaps you're thinking "So what? What does all this mean to me? How can it possibly affect my life—today, tomorrow, in the years to come?"

Well, we could start off by giving you the least selfish reason for caring about the fate of wildlife: Wildlife is important because it is beautiful, worthwhile, and has just as much right to be in the world as humans do. But many people are unwilling to accept this as the only reason. So let's take a look at some of the practical reasons for saving wildlife.

**ECONOMIC AND SCIENTIFIC VALUE** Many products that we use every day and would probably not want to do without are harvested from wild sources. Such valuable products include latex (a rubbery substance used to make balloons, surgical gloves, paint, and other items), wood, and most kinds of seafood.

Many medicines are derived from chemicals extracted from wildlife. Some unusual sources of modern medicines include molds, snake venom, catfish slime, and sponges. Plants too are sources for medicine. In fact, about one fourth of the medicines used today come from plants. One plant that is particularly valuable for its medicinal uses is the rosy periwinkle. Medicines made from this plant are used to treat childhood leukemia (a type of cancer). At one time, only about 1 out of 5 patients with childhood leukemia survived. Now patients are treated with medicines made from the rosy periwinkle and about 19 out of 20 survive. Interestingly, the rosy periwinkle was nearly wiped out when its habitat in the rain forests of Madagascar was destroyed. Can you now explain why medical professionals should be concerned about deforestation?

Wild plants and animals are not only sources of useful products, they are also living banks of information. By studying them, humans can learn about the process of evolution, about the way the body works, and about the nature of behavior. And that's just the beginning!

**GENETIC DIVERSITY** One of the most important scientific reasons wildlife is valuable to humans is that it possesses most of the "library" of genetic information that exists on Earth. This library is made up of units of heredity known as genes. You can think of genes as being extremely short, simple directions. The thousands upon thousands of genes that an organism has work together to determine the characteristics of that organism—what it is, what it looks like, how its body works, and so on.

There are many species of wild animals and plants and the individuals in each species are usually quite different from one another. Thus wild animals and plants have an enormous diversity of genes. This is not the case with domesticated animals and crop plants. Each species of these organisms is made up of a number of varieties, or breeds. Each individual in a particular variety is practically the same as any other individual. For example, a Holstein cow has pretty much the same genes as any other Holstein cow, and a corn plant is almost identical to all the other corn plants in a field. While this sameness has many advantages, it also means that the organisms react to diseases in exactly the same way. What, you might wonder, is the danger in that?

**Figure 4–16** *One Holstein cow looks very much like any other Holstein cow. How does this fact relate to the concept of genetic diversity?*

Suppose a terrible plant disease strikes a corn field. All the corn plants respond to the disease in the same way—they all die. Can you guess what happens when the disease is transmitted to a neighboring field that is planted with the same kind of corn? And what happens when the disease spreads to all the other corn farms in the area, which are also planted with the same kind of corn? That's right—all the corn dies. And the people who were depending on that corn to feed themselves and their livestock are in big trouble.

**Figure 4–17** *In the past, scientists tried to convince South American villagers to "modernize" their agricultural practices. Now, scientists are encouraging them to grow their traditional crops, thus maintaining precious genetic diversity and preserving the villagers' culture.*

Now suppose the same plant disease spreads to a grassy hillside in which some wild relatives of corn are growing. Some of these wild relatives will die from the disease, just as all the corn plants did. But some may have genes that help them to resist the disease. These surviving plants can pass on their genes, including ones for disease resistance, to their offspring. And if scientists know about the genes for disease resistance, they can use techniques of plant breeding or genetic engineering to transfer the genes to domesticated corn plants. The result is the production of corn plants that are resistant to the disease.

**Figure 4–18** *The veterinarians (animal doctors) are doing a checkup on a young whooping crane. This is just one way people can help to preserve the web of life. What are some other ways?*

Finally, imagine the following situation. Long before the outbreak of the corn disease, the hillside is completely cleared of its wild plants and then planted with crops or grass good for livestock. Because of the destruction of their habitat, the wild relatives of corn become extinct. Their genes are lost forever. What do you think will happen when the corn disease strikes? Will scientists be able to develop a disease-resistant variety of corn? Perhaps they will; perhaps they won't. But certainly, their job has become a good deal harder.

**PRESERVING THE WEB** The most selfish reason for caring about the fate of wildlife is also possibly the most compelling. Wildlife is necessary for the continued survival of the human species. In Chapter 1, you learned that Earth's environment can be thought of as a giant spider web. Each thread of the web represents an interaction between a living thing and its living and nonliving surroundings. What happens if too many strands of a spider's web are broken? As Chief Seattle (leader of the American Indian tribes of the Puget Sound area and the person for whom the city of Seattle, Washington is named) noted over a hundred years ago, "Man did not weave the web of life, he is merely a strand in it. Whatever he does to the web he does to himself."

## 4–1 Section Review

1. What is extinction? How do organisms become extinct?
2. What are some human activities that cause organisms to become endangered or extinct?
3. How do human activities affect the rate of extinction?
4. What are deforestation and desertification?
5. Why should people be concerned about wildlife?

**Critical Thinking—*Making Generalizations***

6. Think about the extinct and endangered organisms you have learned about in this section and elsewhere. What sort of characteristics make it more likely for an organism to become extinct or endangered?

# Problem Solving

## Oh, Dear. Deer!

"They ate my prize-winning roses!"
"They ate *my* vegetable garden."
"Did you try planting onions?"
"They ate those, too."
"They killed all the apple trees at the farm down the road last winter. Chewed off the bark."
"I heard one attacked a pickup truck a ways down the road."
"I wouldn't doubt it. The males go crazy in the autumn. Too many hormones or something."
"So much for Bambi."

Many towns in the northeastern United States are under attack—and by of all things, deer! The wolves and mountain lions that preyed on deer and thereby kept their number in check have long been gone from the area, killed off by previous generations of humans. As farms are turned into wooded suburbs, the deer population has skyrocketed.

The imaginary town of Deerfield is desperately searching for a solution to its deer problem. At the moment, the most popular proposal involves importing dingoes, which are wild dogs native to Australia. It is hoped that the dingoes will bring the deer population down.

**Discovering Points of View**

How do you think each of the following townspeople feels about this solution? What alternative solutions might each person propose? Why?

an ecologist
a hunter
a sheep farmer
an animal control officer (dog catcher)
a veterinarian (animal doctor)
an animal-rights activist
a parent with small children
a chemical manufacturer
an apple farmer
an electrician

# 4–2 Seeking Solutions

**Guide for Reading**

*Focus on these questions as you read.*

- *What is wildlife conservation?*
- *What are some methods of wildlife conservation?*

As you have learned in the previous section, human actions can harm Earth's living things. But fortunately for all of us, human actions can also protect them. **The methods used to preserve and protect endangered species, manage populations of wild organisms, and ensure the wise use of living resources are forms of wildlife conservation.** Conservation is the intelligent handling of resources (living

## CAREERS

### Conservationist

The people who see to it that the use of natural resources causes the least possible harm to the environment are called **conservationists.**

There are many positions in the field of conservation. One might work as a soil scientist, range manager, or wildlife manager, to name a few. A conservationist might be involved in water and land management, improving habitats, surveys, or research. People who become conservationists have an interest in nature and have a desire to be part of a program that cares for natural resources. For more information about this rewarding field, write to the Forest Service, U.S. Department of Agriculture, PO Box 2417, Washington, DC 20013.

and nonliving) so that they provide the greatest possible benefit for the longest possible time. Conservation allows us to use part of a resource now and at the same time preserve a sufficient supply of the resource for the future. The conservation of Earth's plants, animals, and other living things is known as **wildlife conservation.**

## Setting Limits

The fate of the dodo and other extinct species has taught us a sad lesson: People cannot be allowed to kill as many living things as they wish. Enough individuals must be left so that a species can maintain its numbers through reproduction. But how can this be achieved? One way of setting limits on the numbers of living things killed is by enacting hunting and fishing laws. Such laws specify how many animals a hunter or fisher is allowed to take from the wild. They may also place restrictions on the species, size, and sex of the animals captured. In addition, hunting and fishing laws specify at which times the animals can be hunted. For example, it is illegal to hunt ducks in the spring and summer, when they are breeding and raising their young.

Unfortunately, these kinds of limits are sometimes not enough. Louisiana's population of American alligators continued to decline rapidly even after the state enacted a law limiting the hunting season to sixty days and allowing each hunter to take only six alligators of a certain size. It was not long before the alligator was in danger of extinction throughout its habitat in the swamps of the southern United States. By 1970, it was necessary to ban alligator hunting in the United States.

Of course, laws need to be enforced if they are to work. And enforcement is often a difficult and dangerous task. Game wardens in the United States, Kenya, Brazil, and elsewhere have been killed in the line of duty. But when the laws are allowed to work, species can be brought back from the edge of extinction. One success story involves the American alligator you just read about. Under protection, the alligators have increased significantly in number. In many areas, they are no longer considered endangered and so they can once again be hunted for

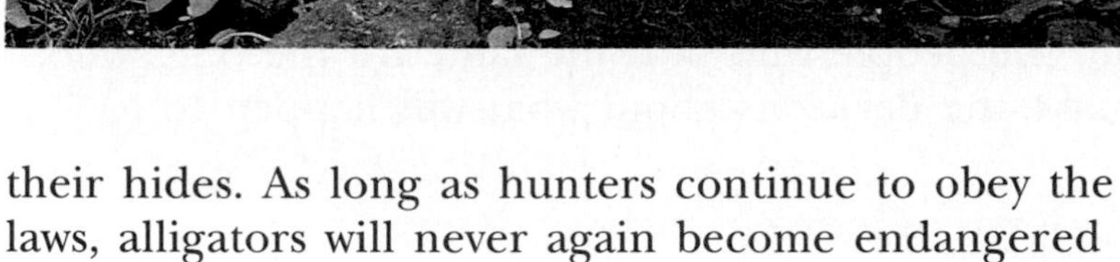

their hides. As long as hunters continue to obey the laws, alligators will never again become endangered from overhunting.

Figure 4–19 *National and international laws protect rare and endangered species. Laws limit the number of American alligators and colobus monkeys that can be killed. Laws also protect the endangered* Rafflesia arnoldi *plant, which has the world's largest flower—more than 90 cm across and 6.8 kg in mass. Unfortunately, the flower smells like rotting meat.*

## Preserving Habitats

You have just arrived at the most important exhibit in the zoo—a parking meter. You cannot believe your eyes. A parking meter? Parking meters belong on streets and in parking lots, not in zoos! What's going on here?

A sign on the parking meter explains it all. The money that people put into the parking meter will be used to purchase land in the tropics. As you hunt through your pockets for change, you recall that habitat destruction has caused more species to become extinct or endangered than overhunting has. Thus, preserving habitats is the most important method of conserving wildlife. Now you agree: It makes sense for the parking meter to be considered the most important exhibit in the zoo.

Have you ever been to Yosemite or any other national park in the United States? If you have, you are probably aware of how hard the National Park Service works to keep the parks as close to their natural state as possible. This helps to preserve wildlife habitats. A few rare or endangered organisms—such as silversword plants, Attwater's greater prairie chicken, and American crocodiles—are found almost entirely in national parks or wildlife refuges. Many

**Figure 4–20** *The parking meter exhibit in the San Francisco Zoo collects donations of spare change. The donations are used to buy and protect wildlife habitats.*

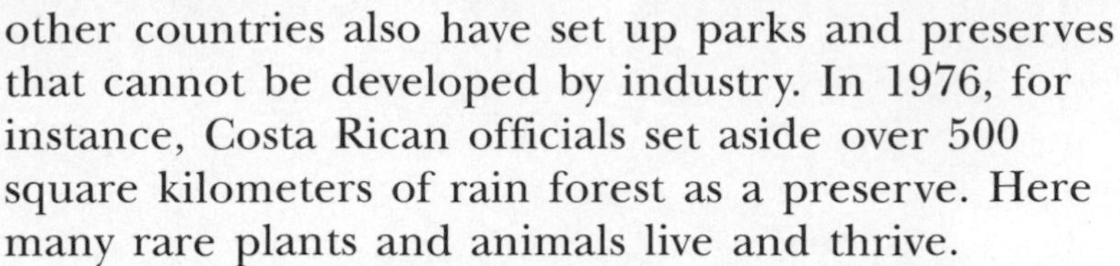

other countries also have set up parks and preserves that cannot be developed by industry. In 1976, for instance, Costa Rican officials set aside over 500 square kilometers of rain forest as a preserve. Here many rare plants and animals live and thrive.

Habitats also need to be preserved in places other than national parks and wildlife refuges. The zoo's parking meter is one way of raising funds for habitat preservation outside of government reserves. By buying land, conservation organizations ensure that the land will remain in its natural state. After all, the people who own the land are the ones who make the decisions about what will happen to it.

**Figure 4–21** *National parks preserve places of natural beauty and wonder. They also protect the wildlife that lives in these places. Yosemite National Park in California is a temperate forest habitat (top). What kind of habitats are found in Everglades National Park in Florida (bottom left) and in Zion National Park in Utah (bottom right)?*

## Raising Reproductive Rates

Sometimes preserving habitats is not enough. To save highly endangered species from extinction, it may be necessary to raise their reproductive rate, or get individuals to produce more offspring. How can this be done?

Reproductive rates are sometimes raised through **captive breeding,** or causing animals in zoos to have offspring. For some species that have vanished from the wild, such as the California condor, captive breeding is the only way of saving the species.

Recently, advanced techniques originally invented for use in humans and cattle have been applied to captive breeding programs. As you can see in Figure 4–22, the results of these techniques are often a bit startling. A "test-tube baby" animal may look quite different from its host mother. But the host mother is usually quite content to care for "her" baby. Test-tube baby techniques allow an endangered-species

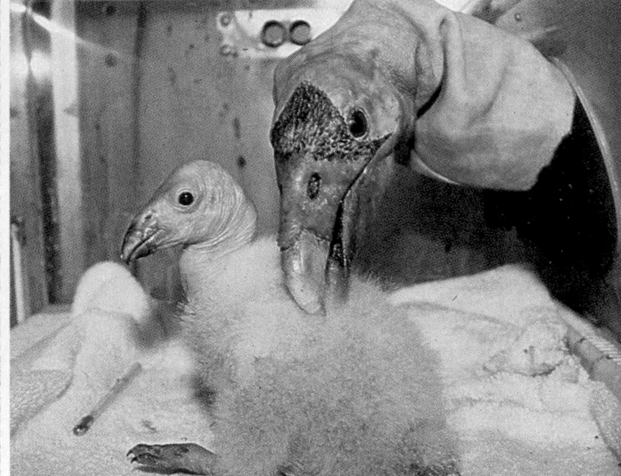

**Figure 4–22** *Because California condors are nearly extinct, it is too risky to allow the real parents to bring up baby. To avoid confusing the condor chick, its keepers use a realistic hand puppet to feed and groom it. Although the test-tube baby bongo does not look much like its eland host mother, the eland takes good care of "her" baby.*

female to produce many more offspring during her lifetime than would otherwise be possible.

Other advanced techniques in reproductive biology allow individuals to produce offspring long after their lifetimes are completed. Scientists are currently developing ways of storing plant seeds and freezing animal sperm, eggs, and embryos so that they can survive in "suspended animation" for many years.

In some cases, captive breeding has been so successful that it has become possible to return animals to their natural habitats. Captive lion tamarin monkeys have been trained to live in the wild, then released into the rain forests of Brazil. In time, the offspring of these monkeys will breed with wild monkeys, thus adding some much-needed genetic diversity to the population. The Arabian oryx, a graceful antelope with long horns, became extinct in the wild in 1972. But because of captive breeding programs started 10 years before in the United States, the oryx was not lost forever to the deserts of the Arabian Peninsula. In the early 1980s, a small herd of formerly captive oryx were released in the country of Oman. Guarded day and night by rangers, the oryxes have thrived. Now, 10 years after their return, there are more than 100 oryx, over three fourths of which have known no home except the deserts of Oman.

## ACTIVITY WRITING

*It's the Law*

The legal protection of living things goes above the state level. One of the most important federal laws (laws that affect the entire United States) is the Endangered Species Act. One of the most important international laws is CITES (Convention on International Trade in Endangered Species of Fauna and Flora).

Using references available in the library, prepare a report on one of these laws.

# CONNECTIONS

## *Computer Dating for Wildlife*

In spite of many cartoons to the contrary, computer dating does not involve machines having a romantic candlelit dinner for two. Computer dating is actually a way of using *computers* to match up potentially compatible people so they can get together socially. For humans, computer dating is a way of meeting people with similar interests. With luck, a person will make new friends and establish new relationships.

For wildlife, computer dating is not a matter of matching up individuals with similar hobbies and outlooks on life. Its purpose is to match up individuals that are as dissimilar as possible in terms of their genes. This helps maintain the genetic diversity of the species. And genetic diversity helps to ensure the continued survival of species, both in zoos and in the wild.

Here's an example of computer dating for wildlife in action. In 1986, 68 European zoos agreed to participate in an international captive breeding program for Siberian tigers. The zoos sent a list of ancestors—parents, grandparents, great-grandparents, and so on—for each of their tigers to an organization that coordinates captive breeding programs. There, information on the zoos' 207 tigers was entered into a computer. The scientists managing the captive breeding program used the computer to analyze the information. The computer's analysis helped the scientists decide which tigers should be bred, when they should be bred, and to which other tigers they should be bred. The scientists also came up with a plan for moving the tigers from one zoo to another in order to meet their computer-assigned dates. So thanks to computer dating, tigers are now off on busy—but romantic—journeys around Europe.

**Figure 4–23** *Captive-bred animals—such as lion tamarins (top left), Arabian oryx (top right), and red wolves (bottom)—have been trained to live in the wild, and then released. In some cases, the animals' release has inspired governments to increase the size of the wildlife preserve in which the animals live.*

## Using People Power

Conserving the Earth's living resources is an awesome task. There are many complex social, political, and ethical issues that affect conservation measures—far too many to be discussed here. But achieving a balance between the needs of humans and the needs of wildlife is not impossible. Indeed, such a balance can ultimately benefit both humans and their fellow passengers on "Spaceship Earth."

In any conservation effort, it is important to keep human needs, attitudes, and desires in mind. When people understand environmental issues and realize how wildlife conservation benefits them, they become some of the best friends wildlife has.

The first step in getting people to support wildlife conservation is to make them aware that there is a problem. Sometimes this is done in spectacular ways: a concert or television special, perhaps. Or an environmental group may host a fair, a fund-raising walk or race, or other event. But the process of making people aware goes on constantly in quieter, less obvious ways. The next time you visit a zoo or park, read a magazine, or listen to an interview of a celebrity—such as Tom Cruise, Meryl Streep, Bette Midler, or Sting—pay attention. You may be

***Figure 4–24*** *Making people aware of conservation issues is sometimes done in spectacular ways. The 1990 Earth Day festival in New York City's Central Park drew thousands of spectators and participants. By placing their bodies and their small boat between whaler's harpoons and whales, Greenpeace activists prevent the killing of whales.*

## ACTIVITY WRITING

*Dare to Care*

Conservation is not just for scientists. All kinds of people from all over the world work together to help save the Earth and its living things. Here are just a few people whose lives and actions have encouraged others to become involved in conservation issues.

George Adamson
Sunderlal Bahuguna
Rachel Carson
Jacques Cousteau
Kuki Gallman
Chico Mendes
Michael Werikhe

Write a brief report on the life and accomplishments of one of these people.

surprised at how many lessons about the environment are being taught.

In many cases, the interests of wildlife and those of the majority of people are not in conflict. Thus people do not have to be coaxed into liking conservation measures. They may already be aware of the problems and anxious to help. But what can ordinary people do?

Quite a lot, once they realize that their opinions matter and they can make a difference. In India, villagers have saved the forests near their homes by hugging the trees and getting in the way of loggers who want to cut the trees down. In Africa, farmers are learning new methods of growing their crops among the trees of the forest. In South America, many Amazon Indian tribes have united to form a powerful political group for rain forest preservation. And there are many, many other examples.

Every person, young or old, rich or poor, can help in the struggle to save Planet Earth. Even little things matter—the little efforts of a lot of people can make a big difference.

Here are a few simple things you can do to help save the Earth and its inhabitants.

- Recycle substances such as paper, steel, glass, and aluminum. This reduces the need for raw materials and helps save habitats from deforestation and mining.

- Write to companies to make them aware of environmental issues. This inspires them to find ways to do less harm to the environment—and even to help it!
- Support companies that are environmentally aware. This encourages them to continue their good work. It may also motivate companies that are less aware to improve their ways.
- Refuse to buy exotic pets, tortoise shell, ivory, furs, and other products made from rare and endangered animals. Ultimately, this makes it less desirable for people to obtain or sell such things.
- Make an effort to think about the ways your actions affect the web of life. This may help you find alternatives that are just as useful to you and more beneficial to the Earth.
- ■ Can you think of other ways to help wildlife and the environment?

Perhaps the most important things you can do to help the Earth and its living things are to learn as much as you can about wildlife and the environment and to share your knowledge with others. Like the Lorax in the famous book by Dr. Seuss, you can "speak for the trees" and the other wild things that "have no tongues" and cannot speak for themselves. You and other people can make your voice the voice of the Earth and its voiceless wildlife. With a good understanding of the past and the present, people can work together to make a better future.

## 4–2 Section Review

1. What is considered to be the most important method of conserving wildlife? Why?

**Connection—*You and Your World***

2. Design a conservation program for the rare or endangered organism of your choice. Explain how your plan takes into consideration the different conservation methods and issues that you read about in Section 4–2.

Paper Route, p.151

*Taking Action*

There are many organizations that are dedicated to preserving planet Earth and its inhabitants. Here are just a few of the larger organizations.

Audubon Society
Greenpeace
World Conservation Union (WCU), formerly International Union for the Conservation of Nature (IUCN)
National/International Wildlife Federation
Nature Conservancy
Sierra Club
Wilderness Society
World Wildlife Fund

Prepare a report on one of these groups. Your report should explain the purpose of the group and describe its activities.

# Laboratory Investigation

## A Miniature World

### Problem

How do human activities affect the environment?

### Materials *(per group)*

| | |
|---|---|
| large jar with cover | 4 aquatic plants |
| table lamp | 8 small pond snails |
| 2 guppies | clean gravel |

### Procedure

1. Place gravel 3 cm deep on the bottom of the jar.
2. Fill the jar with tap water to about 6 cm from the top.
3. Let the jar stand uncovered for at least 48 hours.
4. Using the accompanying diagram as a guide, place plants in the jar.
5. Place the snails and guppies in the jar.
6. Close the jar tightly.
7. Place the jar in a location away from windows and other areas in which temperature and light change greatly.
8. Place the table lamp next to the jar so that the light shines on the jar. The light bulb should be about 15 to 20 cm from the jar.
9. Within 4 to 5 days, the water in the jar should be slightly green in color. If the water does not have any color in it, move the lamp closer to the jar. **Note:** *The light bulb should not touch the jar.* If the water is bright green, move the lamp away from the jar. Adjust the position of the lamp as needed until the water stays a pale green in color.
10. Observe the jar every 2 to 3 days.

### Observations

How did the jar change over time?

### Analysis and Conclusions

1. Why is the lamp necessary?
2. How do the plants and animals in the jar interact?
3. What would happen to the miniature world inside the jar if people killed all the snails and fish?
4. Imagine that an exotic plant disease which killed all the plants was accidentally introduced to your jar. How would this affect the miniature world in the jar?
5. You could have found the answers to questions 3 and 4 by doing something to simulate these forms of environmental damage. Explain why you were not asked to do this.
6. **On Your Own** Design an experiment to test the effects of deforestation on your miniature world. What results would you expect to obtain from your experiment?

## Summarizing Key Concepts

### 4–1 Identifying Problems

▲ A species that no longer exists is said to be extinct. The process by which a species passes out of existence is known as extinction.

▲ Extinction is a natural part of Earth's history.

▲ In the past few hundred years, human activities have greatly increased the rate at which organisms become extinct.

▲ Organisms that are so rare that they are in danger of becoming extinct are said to be endangered.

▲ Some species become rare, endangered, or extinct because they are killed deliberately.

▲ Many species are threatened by the destruction of their habitats.

▲ The destruction of forests is known as deforestation. Deforestation harms both nearby and distant ecosystems.

▲ Although some desertification occurs naturally, most is caused by human activities.

▲ Pollution damages habitats and poses a threat to living things.

▲ Exotic, or non-native, species can upset the balance of interactions in an ecological community and cause native species to become endangered or extinct.

▲ Some activities meant to increase people's appreciation for the natural world may pose a threat to wildlife.

▲ Wildlife has economic and scientific value.

▲ The genetic diversity of wildlife provides keys for solving current and future problems.

▲ Because all living things are interdependent, wildlife is necessary for the continued survival of the human species.

### 4–2 Seeking Solutions

▲ Wildlife conservation is the wise management of Earth's living resources so that they can supply present and future needs.

▲ Preserving habitats is the most important method for conserving wildlife.

▲ In any conservation effort, it is important to keep human needs, attitudes, and desires in mind.

▲ Education is the first step in getting people to help with wildlife conservation.

▲ Captive breeding helps to raise reproductive rates and maintain genetic diversity.

▲ Wildlife conservation is not easy because it is closely linked to many complex social, political, and ethical issues.

▲ If everyone does a little bit to help conserve resources, wildlife and the environment will be helped a lot.

## Reviewing Key Terms

*Define each term in a complete sentence.*

### 4–1 Identifying Problems

extinct
endangered
deforestation
desertification
exotic species

### 4–2 Seeking Solutions

wildlife conservation
captive breeding

# Chapter Review

## Content Review

### Multiple Choice

*Choose the letter of the answer that best completes each statement.*

1. Which of the following is extinct?
   a. rhinoceros  c. snowy egret
   b. dodo  d. lion tamarin
2. The intelligent management of resources is known as
   a. desertification.  c. habitat restoration.
   b. recycling.  d. conservation.
3. Organisms can become endangered because of
   a. habitat destruction.
   b. interactions with exotic species.
   c. overhunting.
   d. all of these.
4. Wildlife is important because it is
   a. a source of valuable products.
   b. a source of genetic diversity.
   c. necessary for environmental balance.
   d. all of these.
5. The process by which a species passes out of existence is known as
   a. endangerment.
   b. extinction.
   c. deforestation.
   d. genetic diversification.
6. A species that is not native to an area is said to be
   a. exotic.  c. extinct.
   b. endangered.  d. endemic.
7. A species that is so rare that it is in danger of disappearing is said to be
   a. exotic.  c. extinct.
   b. endangered.  d. endemic.
8. The destruction of forests is known as
   a. desertification.  c. deforestation.
   b. timber harvesting.  d. defoliation.

### True or False

*If the statement is true, write "true." If it is false, change the underlined word or words to make the statement true.*

1. Extinction is a natural part of Earth's history.
2. The dodo is a(an) endangered species.
3. Enforcing hunting laws is the most important method of wildlife conservation.
4. Elephants have become endangered because of the demand for their meat.
5. Most desertification occurs as the result of natural processes.
6. Wetlands are considered to be extremely valuable ecosystems.
7. The deforestation of greatest concern to people worldwide is occurring in the coniferous forests of the United States.
8. Education does not play an important part in wildlife conservation.

### Concept Mapping

*Complete the following concept map for Section 4–1. Refer to pages G6–G7 to construct a concept map for the entire chapter.*

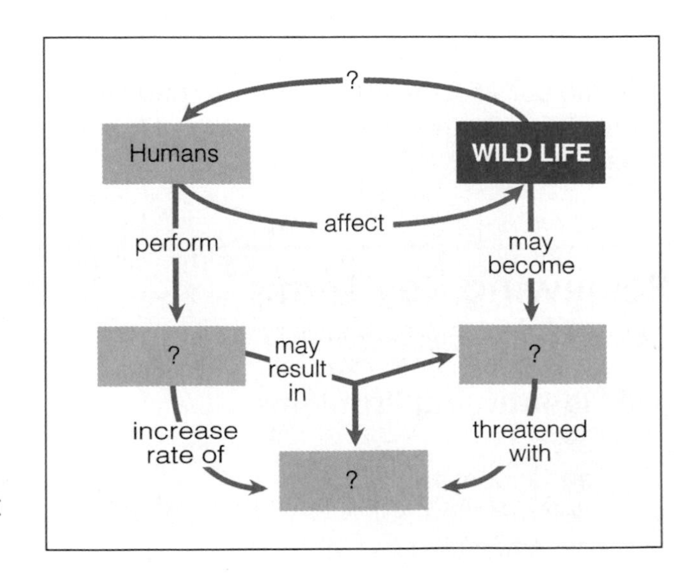

## Concept Mastery

*Discuss each of the following in a brief paragraph.*

1. Explain why it is important to preserve genetic diversity.
2. Discuss five ways in which habitats are damaged or destroyed.
3. What is the difference between the terms extinct and endangered?
4. Giving specific examples, explain how the demands of fashion have caused organisms to become endangered.
5. How do human activities affect the rate of extinction? Why do they have this effect?
6. How does captive breeding help animals in the wild?
7. What does the phrase "loving it to death" mean? How does this apply to the relationship between people and wildlife?
8. List four useful products that come from wildlife.

## Critical Thinking and Problem Solving

*Use the skills you have developed in this chapter to answer each of the following.*

1. **Applying concepts** Suppose one of your friends wanted to buy a pet parrot. What sort of things do you think your friend should know before the parrot is purchased? Do you think buying a pet parrot is a good idea? Why or why not?
2. **Making predictions** In June 1990, Florida passed a law that requires motorboats to go at very slow speeds in certain waterways. In addition, motorboats have been completely banned from a few waterways. How do you think these new restrictions on motorboats will affect manatees?
3. **Relating cause and effect** About 200 years ago, Hawaii's forests were inhabited by 58 species of birds that were not found in any other place in the world. Of these 58 species, 22 are extinct and 20 are endangered. How might each of the following factors have contributed to the disappearance of these birds?
   a. Destruction of three fourths of Hawaii's forests
   b. Exotic species such as sheep, cattle, and rabbits
   c. Exotic species such as cats, rats, and mongooses
   d. Exotic species such as pigeons, sparrows, and doves
   e. Hunting
4. **Using the writing process** Write a short book for first-graders on one of the following topics: endangered species; species that became extinct because of human activities; a method of wildlife conservation. Illustrate your book with your own drawings and/or with pictures from magazines. (Remember to get permission before you cut up the magazines!)

# BIOSPHERE II: A WORLD UNDER GLASS

In the dry foothills of Arizona's Santa Catalina mountains, a privately funded group of scientists has designed and developed a glass-enclosed "terrarium" that contains models of biomes found on Earth. This world in miniature, Biosphere II (the scientists refer to our Earth as Biosphere I), is 1.27 hectares in area, or a little larger than two football fields. It houses some 3800 species of plants as well as fishes, goats, hummingbirds, pigs, moths, bats, and—since early 1991—8 human beings.

The human inhabitants of Biosphere II (who are fondly called Biospherans) will remain sealed in their artificial world for two years. Their goals are: to enhance human understanding of the complex interactions among Earth's living and nonliving things, to develop methods for improved recycling, and to investigate means of establishing self-contained, self-sustaining settlements in outer space. The project is a cooperative effort that has involved many scientists from different fields.

Under its glass dome, Biosphere II is divided into several parts. In one corner stands a five-story building complex that contains apartments as well as research, recreational, and health facilities. Next to the building complex is a farm that is approximately the size of an ice-hockey rink. At the farm, grains, garden vegetables, and tropical fruits are grown in soil enriched with manure, compost, and earthworms. Livestock and freshwater fishes are also raised on the farm.

▲ **From the top of Biosphere II's mountain, you can see the marshes, (top left), ocean (center left), savanna (top right), and rain forest and its stream (bottom).**

Alongside the farm and building complex, the remaining environments lie in a carefully planned sequence. Next to the farm is the model rain-forest biome. The glass dome above the rain forest soars 26 meters above the forest floor (about the height of a seven-story building) to house a mountain, a forest, a pond, and 300 species of plants. The rain-forest plants were selected by Dr. Ghillean Prance, director of the Kew Royal Botanic Gardens in London, England. From this model, scientists hope to learn ways to rebuild damaged rain forests.

Water from the rain forest flows down the miniature mountain and collects in a stream that runs through the next model biome, the savanna. This model is the joint effort of Dr. Tony Burgess, an ecologist and botanist who runs the U.S. Geological Survey's Sonoran desert lab, and Dr. Peter Warshall, a biologist and anthropologist. The savanna contains microbes, insects, and hummingbirds as well as species of grasses from Africa, Australia, and South America.

Next, the stream runs into a small freshwater marsh then ends its journey in a saltwater marsh. Dr. Walter Adey, a marine biologist at the Smithsonian Institute, designed these marshes in addition to the lagoon, coral reef, and ocean. The marine habitats are homes for fishes, shellfish, corals, and sea plants.

Beyond the miniature sea is Biosphere II's final model biome—a cool, foggy desert designed after Mexico's Baja California. Dr. Burgess and Dr. Warshall (who also designed the savanna) chose the 300 plant species and animals for this area, where cacti and mesquite bushes share space with kangaroo rats, insects, and even scorpions.

Biosphere II works as a self-contained unit. Plants take human and animal wastes and turn them into the oxygen and food needed for animal life. Air is purified by pumping it through the farm's soil, where microorganisms strip it of pollutants. Water is purified through the interactions of plants, fishes, and microorganisms in the farm's fish tanks. (Drinking water is further purified through evaporation and condensation.) Human waste is treated to kill harmful microorganisms, then flows into the marshland, where microorganisms and water hyacinths break it down and purify it.

In spite of all this careful planning, the scientists working on the project admit that there are many uncertainties in Biosphere II. So beneath Biosphere II lie pipes and pumps and fans designed to cool and filter the air, store water, recycle wastes, and move the structure's air. Biosphere II also has a power plant to provide the project with electricity.

Despite its uncertainties, Biosphere II offers scientists a unique research opportunity. In Biosphere II they are discovering ways to improve the Earth's enviroments as well as ways for humans to survive far from it in outer space.

▲ **The eight Biospherans have an enormous responsibility: caring for an entire world.**

# Conservationists to the Rescue?

Bathed in early morning light, a herd of large, dark brown sambar deer grazes on a grassy hillside at the edge of a forest in southern Asia. Suddenly, one of the deer raises its head in alarm and sniffs the air. Soon the entire herd is wary and watchful–ready to flee in an instant.

The reason for their alarm quickly becomes obvious. Emerging from the woods a mere 225 meters away, a tiger strides up the hillside. The big cat's powerful muscles ripple as it moves. For a moment, the tiger stops and turns its head toward the deer. Then it continues on its way. This morning, at least, the tiger is not hungry. The deer are left to graze in peace.

Tigers are a vanishing species. They have disappeared from many parts of Asia where they were once common. In an effort to save them, conservation groups such as the World Wildlife Fund have raised millions of dollars. This money is used to support research into tiger behavior and to establish preserves in the wild where the cats are protected.

To some extent, the effort has been successful. India now has 11 preserves for tigers. In some of these protected areas, the cat population is increasing once more. Conservationists are hopeful that tigers can continue to live in the wild.

But not everyone wants tigers living in the wild. In areas near some of India's preserves, tigers have killed both animals and people. Some Indians living near preserves oppose the government's attempts to save the tiger.

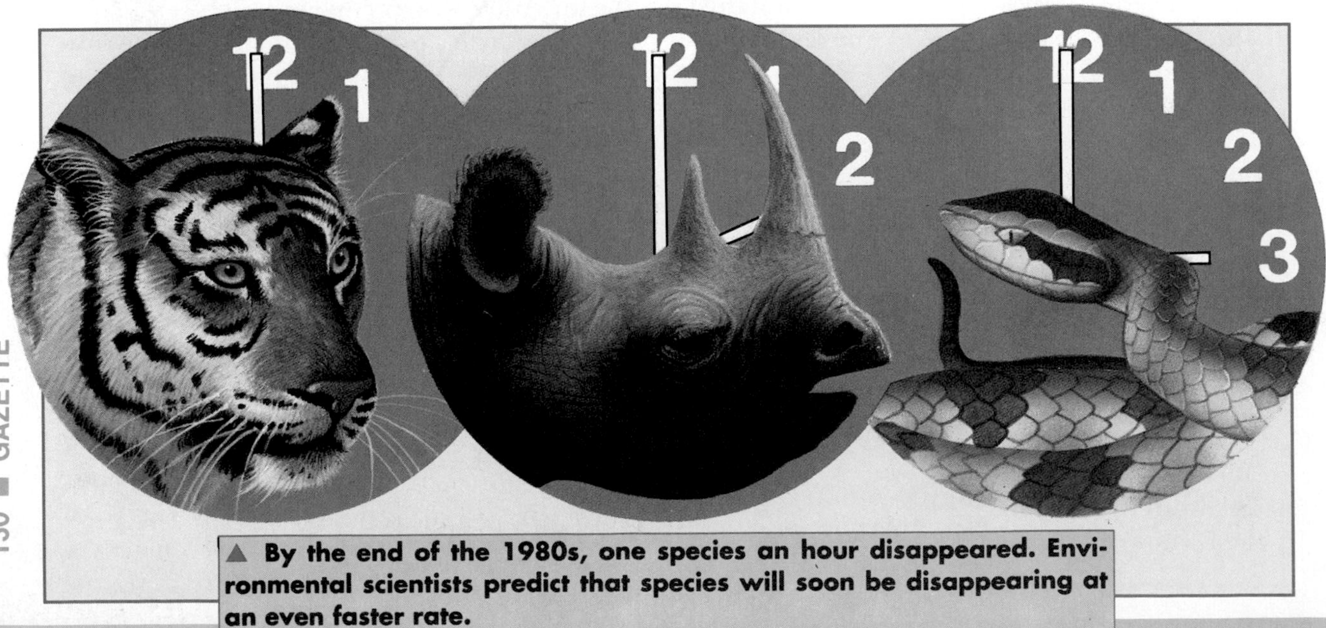


▲ **By the end of the 1980s, one species an hour disappeared. Environmental scientists predict that species will soon be disappearing at an even faster rate.**

They feel the tiger is not an endangered species that should be saved, but a dangerous threat that should be removed.

## A FIGHT TO THE FINISH

Each year, more plants and animals become endangered. As conservationists work to save them, questions often arise about whether all species of living things should be saved. And if not all, which ones should be saved?

The answers are not simple. Some situations involve a conflict between the needs of humans and the needs of wildlife. Other situations require a decision about which species are worth time, effort, and money. For example, should developers be prevented from building in the Pine Barrens forest of New Jersey if the construction endangers a type of moth that lives there? Or is it worth thousands of dollars to save a kind of sparrow that lives only in a small part of Florida near Cape Canaveral?

## FACING THE FACTS

To do their job, conservationists have to face tough questions like these. Some leading conservationists are very practical about the issues. Norman Meyers, for example, is a well-known environmental scientist. He notes that each hour another species, counting insects and other invertebrates (animals without backbones), disappears.

"Sad to say," Meyers writes, "the question is not how to save *all* these species; we just do not have the resources to rescue more than a small fraction."

Meyers suggests that first the value of each species be determined. The decision would be based on economics, the environment, and the survival of people. Then, the most valuable species should receive help first.

For example, in the African country of Kenya, lions are a big tourist attraction. Tourism brings lots of money to the country. Therefore, lions may be worth saving.

▶ **This California condor is one of many species of animals that is threatened with extinction.**

A tropical plant called the periwinkle is the source of two drugs used to treat cancer. And the venom of the Malayan pit viper, a cousin of the rattlesnake, is used to stop blood clots that cause heart attacks. In Meyers' view, the periwinkle and the pit viper should be among the first to receive help.

## THE UNKNOWN FACTOR

It is clearly to our advantage to invest time and money in preserving useful species of plants and animals. The trouble is, there are probably thousands of species that could be very valuable to people–only no one yet knows it!

Are we wiping out species that hold secrets that would help us cure cancer and AIDS, feed a hungry world, or solve the energy crisis? Scientists and conservationists Paul and Anne Ehrlich of Stanford University think we might be. And they also see a further complication. When species disappear, the ecosystems, or environments, to which they belong are changed or even destroyed. All life on the Earth depends on ecosystems. Ecosystems provide important services such as the maintenance of soils and the control of crop pests and transmitters of human disease. As the Ehrlichs write, "Humanity has no way of replacing these free services should they be lost... and civilization cannot persist without them."

The Ehrlichs cite yet another reason for saving as many species as possible–"plain old-fashioned compassion." For many conservationists, stopping the extinction of species is a moral issue.

## CONCERN AROUND THE WORLD

The International Union for the Conservation of Nature and Natural Resources (IUCN) is a worldwide group associated with the United Nations. The IUCN's position is that we are "morally obliged to our descendants and to other creatures" to act wisely when it comes to conserving plants and animals.

Most conservationists would agree that each time a species vanishes, the world is a bit poorer. Consider just the sheer beauty of many of the endangered plants and animals. The main reason for saving tigers and condors, for example, is that they are among the Earth's most magnificent creatures.

Perhaps sadly, people tend to feel more concern for species that are pretty or striking than for those that are plain. The black rhinoceros of Africa and the Higgin's eye mussel of rivers of the Midwest are both in danger. The world knows about the threat to the rhinoceros. But few people know about the Higgin's eye, even though it is closer to extinction than the rhino.

How can conservationists decide which endangered species to help or, at least, to help first? Norman Meyers has an interesting approach. When many people are injured in a big disaster, physicians often first treat victims who are badly hurt but who will recover if treated promptly. Meyers suggests that the same approach be used for species in danger. First help those species that are in the greatest danger and that have a good chance of surviving extinction. Still, says Meyers, it will not be easy to decide which species to aid and which to ignore. Making these decisions, he adds, "will cause us many a sleepless night."

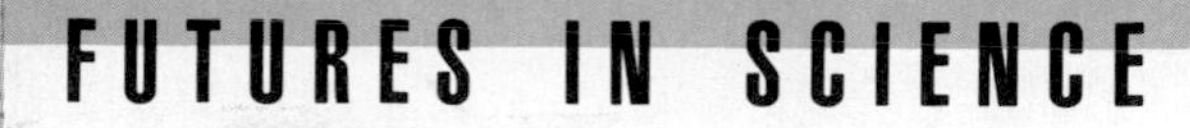


# What Became of Africa's Animals?

The sun is not yet up. In the kitchen of a comfortable home on the outskirts of a medium-sized city, a man is brewing coffee. Looking at a calendar, he checks the date—January 24, 2050. Although it is still cool outside, the man wears short pants and a short-sleeved shirt. He expects another scorching day here in East Africa.

Just as dawn breaks the man jumps into a vehicle and drives away from the city. On each side of the road sprawl rich farms with vast, neatly cultivated fields of crops. Within 20 minutes, he pulls up to a gate, where a guard greets him. His workday has begun. This man is a game warden at a large national park and wildlife sanctuary.

At the same time every morning, the warden makes the rounds to see how the wild animals in the park are doing. The road over which he drives is paved with blacktop. The warden wonders how it must have been back in the 1990s, when wardens drove battered cars over roads that were nothing more than dirt tracks.

At that time, the park was far away from any farms and cities. All around it stretched grassy plains. Wild animals wandered freely inside and outside the park. In 2010, however, the land near the park was set aside for farming. The population of the country had grown so large that all the fertile land was needed to produce food. Only the parks were left wild. "At least," the warden thinks to himself, "my country saved the parks." In a country nearby, the need for food had been so great that even the national parks were turned into farms.

As the warden reaches the park's western boundary, he stops to check the condition of a high wire fence that encloses it. The fence was built 20 years earlier to stop animals from wandering off park property onto farmland. Elephants, buffaloes, antelopes, and zebras were eating the farmers' crops. Lions were killing livestock and threatening people. When the farmers complained, the fence was put up.

The fence was not strong enough to stop the elephants, however. They broke through the fence and kept destroying crops. In the end, there was only one solution to the problem. The elephants had to go. Most were killed, but some young ones were caught and shipped to a special park far away from both farms and cities. It is one of the few parks that are big enough to contain such large, far-roaming animals. Only these parks still have elephants and, for that matter, lions. Even with fences separating them, people do not want to live next door to animals as dangerous as lions. So lions are permitted only in the most distant parks.

## KEEPING ANIMALS ALIVE

Back in his vehicle, the warden drives through a broad valley. On all sides are antelopes and zebras, even a few buffaloes. The warden stops at a large concrete drinking tub. He switches on a pump and watches the tub fill with water. It is the dry season in East Africa, and the natural watering holes in the park have dried up. Water must be pumped out of the ground for the animals to drink. Before the fence went up, they could leave the park during the dry season to drink at rivers out on the plains. But the animals no longer have that freedom, so the warden must provide water for them.

As the warden continues his rounds, he meets two of his park rangers. They are counting the number of wildebeests, a type of antelope, in a herd. Years ago, more than a million wildebeests visited the park each year in the wet season. When the parkland

**▶ Wildebeests travel the African countryside in search of green grass to eat. During the dry seasons, the wildebeests must be able to leave an area with dried grass for an area with moist grass. If the human population continues to grow, however, the wildebeests will be unable to roam great distances. What will the consequences be?**

dried out and the grass turned brown in the dry season, the wildebeests would leave. They would travel to an area 160 kilometers away to find water and green grass to eat. But that area was finally taken over by farms and villages. Because they no longer had a place to go in the dry season, most of the wildebeests died. The government's conservation department rounded up the rest of the wildebeests and drove them back to the park. They can survive in the park because of the drinking tubs. But the small amount of grass left on the land during the dry season can support few wildebeests.

So the number of wildebeests must be kept below 20,000 or they will starve. That is why the rangers keep count. Once the herd grows larger than 20,000, the warden and his rangers have to kill some of the older animals to keep the size of the herd down. The meat is given to the farmers living just outside the park.

The warden looks over the wildebeest herd with pride. True, it is not nearly as impressive as the herds that used to roam 40 to 50 years ago, he thinks. But still people have a chance to see the animals. The warden remembers how, on a trip to the United States, he visited a national park in the western part of that country. He saw shaggy American bison walking around in the park. Two centuries had passed since the bison wandered freely far and wide, but in the parks they had been preserved. It is the same with his country's wildebeests.

## TWILIGHT THOUGHTS

Late in the day, the warden pauses at a rocky hillside. For a moment, he glimpses a flash of yellow fur with black spots. Then it is gone. "A leopard," he whispers to himself. It has been a year since he has seen one, although he knows that at least six leopards inhabit the park.

Leopards, unlike lions, are meat eaters that seem able to survive near large numbers of people. Leopards move about mostly at night and are loners. Lions are active partly in daylight and live in groups. It takes the meat of many large animals, such as wildebeests and zebras, to feed a group of lions. But a leopard can live very well on smaller prey, such as baboons.

Driving home, the warden thinks about how wonderful it is to be able to see leopards prowl and zebras roam. He cannot help thinking that if his country had fewer people, there could be more wildlife parks. "But," he reminds himself, "we've done our best."

# For Further Reading

If you have been intrigued by the concepts examined in this textbook, you may also be interested in the ways fellow thinkers—novelists, poets, essayists, as well as scientists—have imaginatively explored the same ideas.

**Chapter 1:** Interactions Among Living Things

Attenborough, David. *The Trials of Life.* Boston, MA: Little, Brown.

Defoe, Daniel. *Robinson Crusoe.* New York: Penguin.

London, Jack. *Call of the Wild.* New York: Macmillan.

**Chapter 2:** Cycles in Nature

Adams, Richard. *Watership Down.* New York: Avon.

Burton, Jane, and Kim Taylor. *Nightwatch.* New York: Facts on File.

Kipling, Rudyard. *The Jungle Book.* New York: Penguin.

**Chapter 3:** Exploring Earth's Biomes

Carson, Rachel. *The Edge of the Sea.* New York: Signet.

George, Jean C. *Julie of the Wolves.* New York: Harper & Row Junior Books.

George, Jean C. *My Side of the Mountain.* New York: Dutton.

Lopez, Barry. *Arctic Dreams: Imagination and Desire in a Northern Landscape.* New York: Scribner.

Perry, Donald. *Life Above the Jungle Floor: A Biologist Explores a Strange and Hidden Treetop World.* New York: Simon and Schuster.

**Chapter 4:** Wildlife Conservation

Henry, Marguerite. *Mustang, Wild Spirit of the West.* New York: Macmillan.

Lasky, Kathryn. *Home Free.* New York: Macmillan.

Mowat, Farley. *Never Cry Wolf.* New York: Bantam Books.

# Activity Bank

Welcome to the Activity Bank! This is an exciting and enjoyable part of your science textbook. By using the Activity Bank you will have the chance to make a variety of interesting and different observations about science. The best thing about the Activity Bank is that you and your classmates will become the detectives, and as with any investigation you will have to sort through information to find the truth. There will be many twists and turns along the way, some surprises and disappointments too. So always remember to keep an open mind, ask lots of questions, and have fun learning about science.

# GARBAGE IN THE GARDEN

More and more people are starting to put garbage in their gardens. You might think this practice would endanger health, smell bad, and damage the garden. But if it is done properly, it is safe, free of unpleasant odors, and beneficial to plant growth. Thanks to the action of helpful bacteria (which are often assisted by burrowing creatures such as worms), the garbage breaks down to form a dark-colored, nutrient-rich substance called compost.

In this activity you will explore the process of making compost.

## Materials

- 2-L clear plastic soda bottle
- small nail or push pin
- scissors
- topsoil
- 150 mL beaker
- scraps of paper
- grass clippings
- weeds and leaves
- uncooked vegetables and fruit scraps
- china marker
- cheesecloth
- rubber band
- plastic fork
- foam meat tray

## Procedure

1. Carefully poke holes in the sides and bottom of the soda bottle with the nail. Use the accompanying diagram as a guide. **CAUTION:** *Be very careful and take your time.* Take turns making the holes—there are many to make.

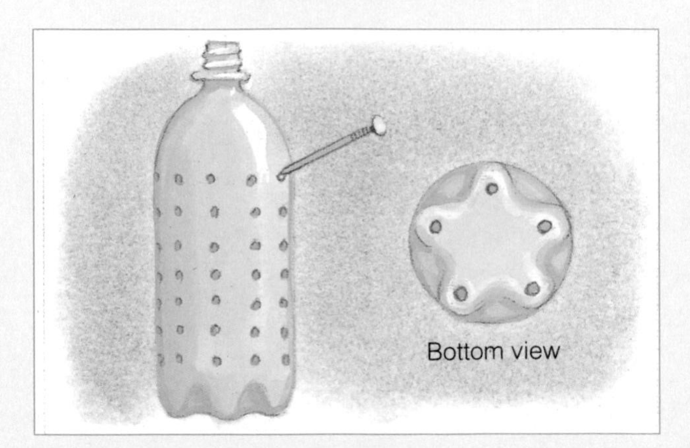


2. Using the nail, poke a large hole near the top of the bottle at the point where the sides become vertical and the plastic thins out. Starting at this hole, carefully cut off the top of the bottle with the scissors. **CAUTION:** *Be careful when working with sharp objects.*

3. Label your bottle with the names of the people in your group and the date. Then put the bottle on the meat tray.
4. Fill the bottle about one-third full with grass clippings. Add 100 mL of soil, then 20 mL of water. The contents of the bottle should be moist but not soaking wet. If the contents are still dry, add a little more water.
5. With the scissors, cut the paper, leaves, weeds, and vegetable and fruit scraps into pieces no larger than 1 cm across. Fill the bottle about one-half full with the cut-up materials.

6. Use the plastic fork to mix the contents of the bottle well. If any materials fall onto the meat tray, lift the bottle, remove the tray, and dump the tray's contents into the bottle. Then put the tray back under the bottle.
7. Make a mark on the outside of the bottle to indicate the height of the contents. Cover the bottle with a piece of cheesecloth. Secure the cheesecloth with the rubber band. Then place the bottle in a warm location.
8. Twice a week, observe the contents of the bottle. Touch the sides of the bottle and note whether the bottle feels warm or cool to the touch. **CAUTION:** *If the bottle feels hot, notify your teacher immediately.*
9. After you have made your observations, add some water if the contents are dry. Once a week, mix the contents with the fork and mark their height on the side of the bottle. Write the date next to the new mark. Make sure you replace the cheesecloth when you have finished making your observations. **Note:** *If the contents start to smell bad, like rotten eggs or vinegar, mix the contents every time you make your observations.*

*(continued)*

## Observations

1. How did the contents of the bottle change over time?
2. What kind of fruit and vegetable scraps did you add to the bottle? Which kinds of scraps decomposed the fastest? The slowest?
3. How did the level of the bottle's contents change over time?
4. How long did it take for the contents of the bottle to finish decomposing?
5. What does your "finished" compost look like?

## Analysis and Conclusions

1. Why did the materials in the bottle change?
2. Why do you think you had to put holes in the bottle and cover the bottle with cheesecloth?
3. Some decomposer bacteria use oxygen, a substance that makes up about 21 percent of the air you breathe. Others do not use oxygen, but are not harmed by it. Still others are slowed down or even killed by oxygen. What can you infer about decomposer bacteria and making compost?
4. If the materials in a compost heap are not mixed regularly, it may start to smell bad. Explain why this might occur.
5. Predict what would have happened if the bottle and the materials in it had been sterilized. Would your results have been the same?
6. Compost improves the texture of garden soil, making it easier for plants to grow in it. It also acts like fertilizer. Explain why compost adds nutrients to the soil.
7. Landfills—commonly known as garbage dumps—are running out of space. In some parts of the country, little space is available for making new landfills. And people desperately need landfill space to dispose of garbage. How might making compost solve part of the landfill crisis?
8. Imagine that you and the members of your group have just been selected as part of a special task force. Your job is to devise a plan that will reduce by 90 percent the amount of yard waste that is ending up in the local landfill. Keep in mind the needs, attitudes, and abilities of the different kinds of people in your community as you work on the plan.

## Going Further

Design an experiment to test how the formation of compost is affected by one of the following factors: light, heat, moisture, air.

# ON YOUR MARK, GET SET, GROW!

In every environment, the supply of resources is limited. In fact, some resources are so limited that there is not enough to go around. Because of this, organisms must compete with one another for the scarce resources. Competition can occur between species and within a species. How do different degrees of competition affect the growth and survival of organisms? Find out in this activity.

## What Do You Do?

1. Obtain four clean half-gallon milk cartons, dried beans, soil, scissors, and a large plastic or foil tray.
2. Using the scissors, cut off the side that has the opened spout on each of the milk cartons. With the pointed tip of the scissors, poke three or four holes in the opposite side.
3. Fill each carton about two-thirds full of moist soil. Using your finger to make the planting holes, plant beans 3 cm deep and 8 cm apart in the first carton. Plant beans 3 cm deep and 5 cm apart in the second carton, 3 cm apart in the third carton, and 1 cm apart in the fourth carton.
4. Place the cartons side by side on the tray. Put the tray in a sunny spot.
5. Water the soil regularly so that it remains moist, but not soaking wet. Observe the cartons daily. Record your observations.

## What Did You Learn?

1. Did the seeds sprout at the same time? Did the seedlings grow at the same rate and in the same way? Describe how the seedlings grew.
2. Did the seedlings in some cartons grow better than the seedlings in other cartons? Why do you think this was the case?
3. Compare your results with those of your classmates. Did you obtain similar results? Why do you think this was the case?
4. What are some possible sources of error in this activity?
5. Using what you learned in this activity, explain why gardeners have to thin out seedlings.
6. Weeding makes gardens look nicer. But is there any other reason for weeding? Explain.
7. What effect(s) do you think fertilizer would have on the space needs of bean seedlings?

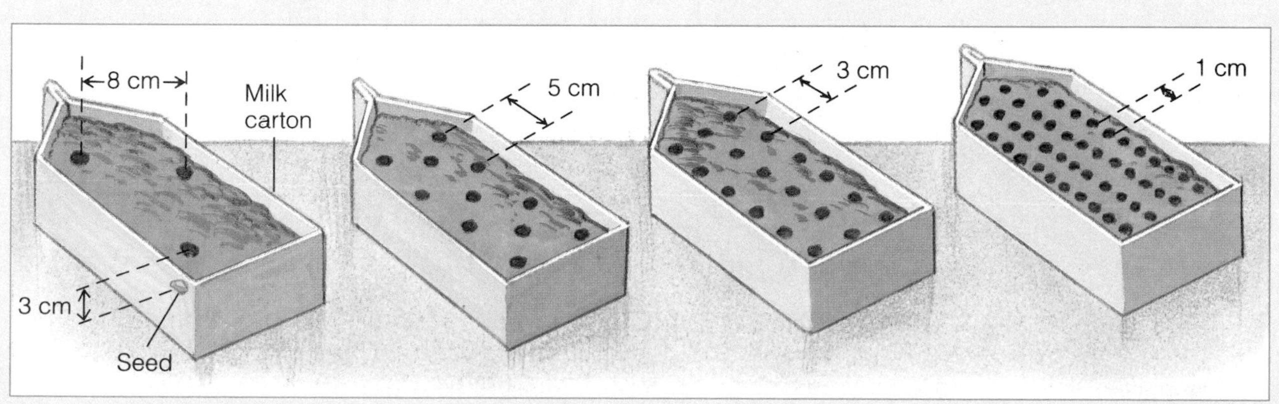

# GOING THROUGH A PHASE

Did you ever notice that the moon looks different at different times of the month? One night, it's just a sliver of light. Ten nights later, it's a big round ball of light. And over the course of the next fourteen nights, it gradually shrinks until there seems to be no moon at all!

The moon goes through its endless cycle of growing and shrinking because it revolves around the Earth, which in turn revolves around the sun. The relative positions of the Earth, moon, and sun determine how much of the moon's surface reflects light back to the Earth. How? Find out in this activity.

## What Do I Need?

10–15 cm ball of Styrofoam or yarn

knitting needle or dowel about 30 cm long and 0.5 cm in diameter

room with lamp

## What Do I Do?

1. Carefully stick the point of the knitting needle into the ball. The ball represents the moon.

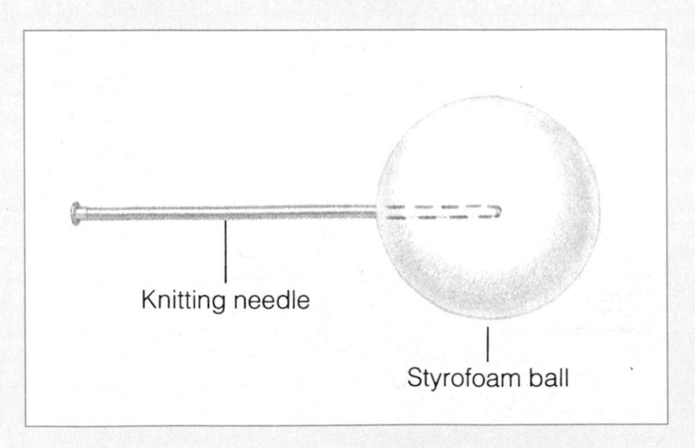


2. Turn on the lamp. The lamp represents the sun. Turn off any other lights in the room and close the shades if it is still daytime.
3. Stand in the middle of the room, facing the lamp. You represent an observer on Earth. Hold the knitting needle at arm's length in front of you. How much of the part of the "moon" facing you is illuminated by the lamp? How much is in darkness? Record your observations and make a sketch to show what the moon looks like.
4. Turn your body 45° to the left. (A complete circle is 360°; an "about-face" is 180°.) Keep your arm and the ball you are holding in the same position. Record your observations and make a sketch of the moon.

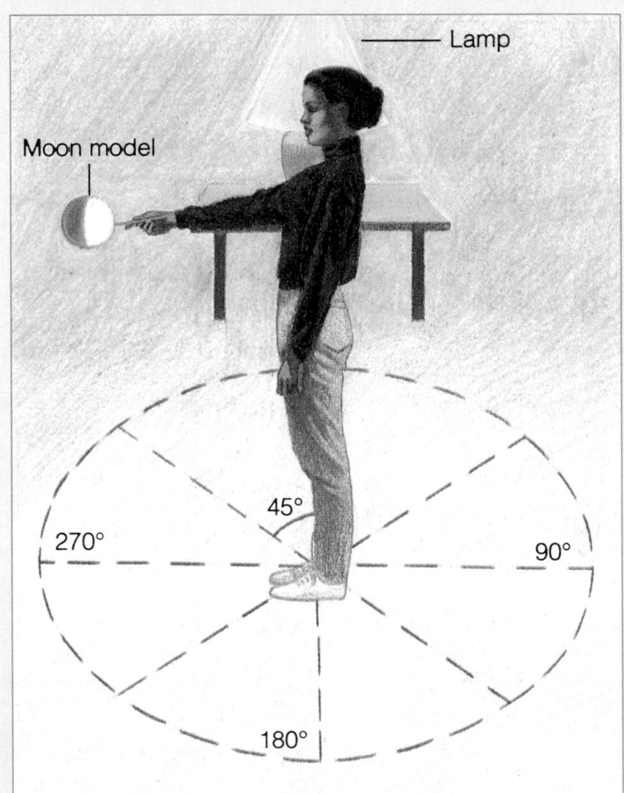


5. Repeat step 4 six more times. One final turn of 45° should bring you back to where you started.

## What Did I Learn?

1. During a lunar cycle, the moon appears to change shape. These shapes are known as phases. In this activity you modeled all of the phases. How many phases are there?
2. The phase in which the moon is not visible in the sky is called the new moon. Describe the position of the Earth, sun, and moon during the new moon.
3. When the moon and the sun are at right angles to one another (270° and 90° in the circle), the high tides are at their lowest. Describe how the moon appears during these lowest high tides.
4. When the sun, moon, and Earth are in a straight line, the high tides are at their highest. Describe the appearance of the moon during the highest high tides.
5. Review page G51. What biological events are associated with the highest high tides? With the lowest high tides?
6. Why is it critical for the biological clocks in grunion to be in harmony with the phases of the moon? (*Hint:* You may find it helpful to review pages G45 and G51.)
7. Prepare a brief illustrated report or a poster to share your findings with your classmates.

# A SAUCEPAN SIMULATION OF A CYCLE

The processes in the water cycle continuously move water between the Earth's surface and the atmosphere. In this activity you will observe the basic processes that form the water cycle: condensation, evaporation, and precipitation.

## Materials

- 2 same-sized jars (such as those used for baby food), one of which has a lid
- 2 same-sized flat containers such as pie pans
- measuring cup or 250-mL beaker
- medium-sized saucepan (about 2.5 L, or 2 qt)
- hot plate or stove
- small saucepan with long heat-proof handle (about 1.5 L, or 1 qt)
- ice
- oven mitt or heat-proof glove

## Procedure

1. Measure 100 mL of water into each jar and each flat container. Securely cover the appropriate jar with its lid. Put the jars and one of the flat containers in a sunny place. Put the other flat container in a dark closet.
2. Let the jars and the containers stand for about a day, then examine them. Use the measuring cup to measure the amount of water in the jars and containers. Record your observations.
3. Fill the large saucepan about one-fourth full with water. Put the pan on the hot plate. Turn on the hot plate and bring the water in the pan to a boil. **CAUTION:** *Be careful when working with a heat source.* Record your observations.
4. Fill the small saucepan with ice cubes. Put on the oven mitt. Grasp the handle of the small saucepan with your covered hand. Hold the small saucepan so that the main part of the pan, but not the handle, is over the boiling water in the large saucepan. **CAUTION:** *Do not put your hand or any other part of your body directly above boiling water. Steam can cause bad burns.*
5. Watch what happens to the bottom of the small saucepan. Record your observations.
6. Turn off the hot plate and clean up after your equipment has cooled off.

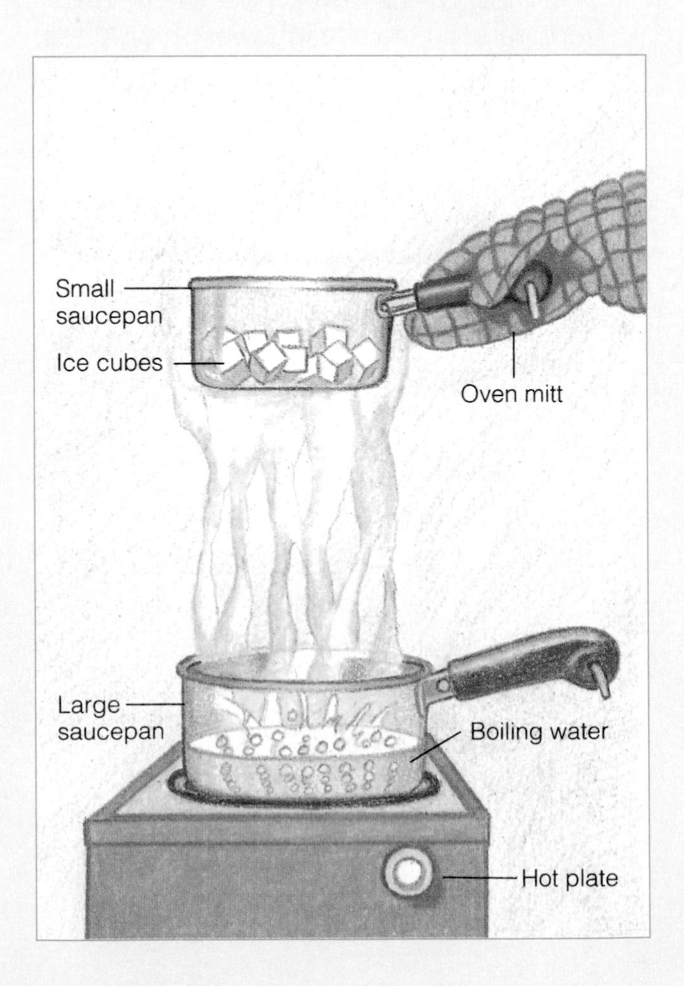

## Observations

1. What happened to the water in the jars and flat containers?
2. Describe what you observed about the boiling water.
3. What happened to the bottom of the small saucepan?

## Analysis and Conclusions

1. What water-cycle process caused the results in step 2?
2. Using the results from step 2, compare the amount of water in the two jars, in the two flat containers, and in the open jar and the flat container that were in the sunny place. What can you conclude about evaporation from these results?
3. Identify the water-cycle processes that you observed in step 5. Explain your answer.
4. Compare your results with those of your classmates. Are they similar or different? Explain why.
5. On a hot, sticky summer day, you notice water collecting on the outside of your glass of ice water. One of your friends says that the hot weather has caused the pores in the glass to open up so that the water leaks out. You suspect that the water got on the outside of the glass another way. What is your hypothesis? How might you go about testing your hypothesis?
6. Cold air can hold less water than warm air. The loss of enough heat energy causes water vapor to condense into liquid water. Use this information to explain the following events.
   - **a.** Steam condenses on a pan that contains ice.
   - **b.** Dew forms on grass during the night.
   - **c.** Morning fog disappears as the day gets warmer.

   How do these events relate to the water cycle?

# TAKE IT WITH A GRAIN OF SALT

Have you ever seen pictures of the stone spikes, columns, and other strange formations inside a cave? These unusual and often beautiful rock formations, like the "castles" of salts in Mono Lake (shown on page G37), are byproducts of the water cycle. Discover what the water cycle has to do with rocks in this activity.

## What You Need

2 small (about 235 mL) drinking glasses or jars
Epsom salts
spoon
saucer or jar lid
30 cm of yarn or heavy cotton string
20 cm of thread or dental floss
pencil

## What You Do

1. Assign roles to each member of the group. Possible roles include: Recorder (the person who records observations and coordinates the group's presentation of results), Materials Manager (the person who obtains all materials and coordinates cleanup), Principal Investigator (the person who reads instructions to the group, makes sure that the proper procedure is being followed, and asks questions of the teacher on behalf of the group), and Specialists (people who perform specific tasks such as preparing the glasses of solution, setting up the yarn part of the experiment, or setting up the thread part of the experiment). Your group may divide up the tasks differently, and individuals may have more than one role, depending on the size of your group.
2. Fill the glasses about three-fourths full with water. Add Epsom salts to the water a teaspoon at a time and stir. Keep adding Epsom salts to the water until no more will dissolve.
3. Place the glasses so that the saucer is between them. Then place the ends of the yarn in the glasses, as shown in the accompanying illustration. Note that the center of the yarn is lower than the ends and hangs over the saucer without touching it.

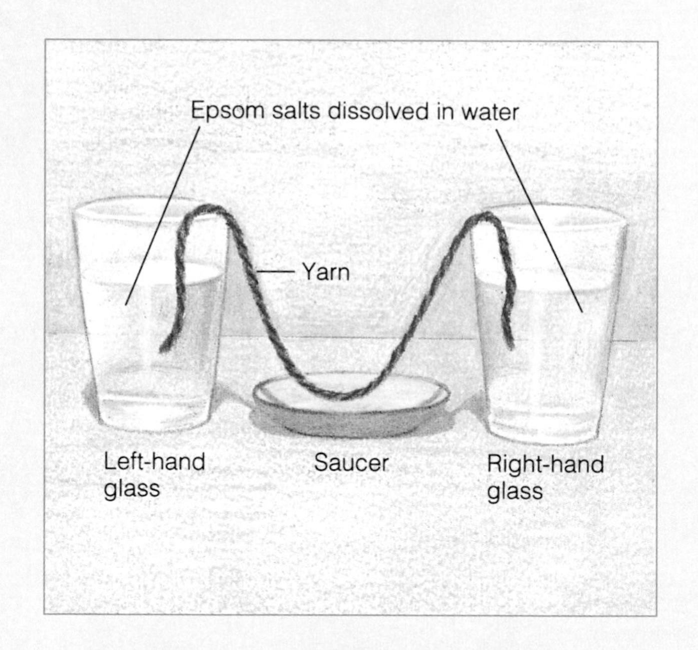


4. Tie a knot in one end of the piece of thread. Tie the other end of the thread around the pencil. Adjust the length of the dangling part of the thread so that knot will hang about 1 cm above the bottom of the left-hand glass when you rest the pencil across the top of the glass.
5. Moisten the knot with water and dip it in the Epsom salts. Then rest the pencil

across the top of the left-hand glass. Make sure that the knotted thread is submerged.

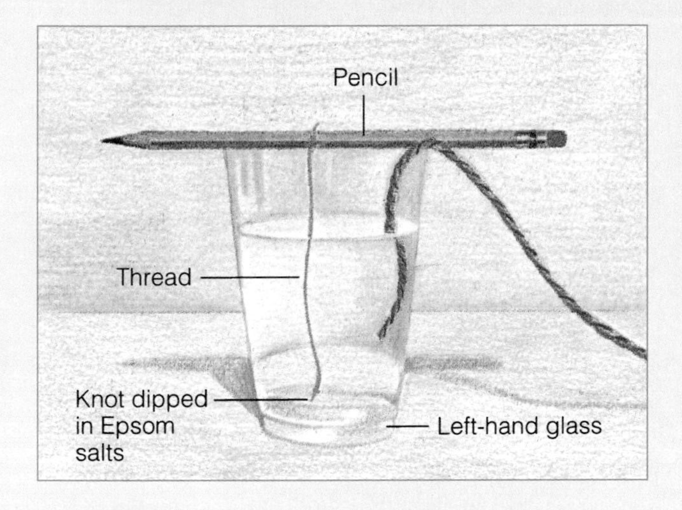


6. Observe the glasses, saucer, yarn, and thread daily. Record your observations.

## What Did You Learn?

1. What do you think is the purpose of the saucer?
2. What happened to the glasses? Why do you think this occurred?
3. What happened to the thread?
4. What happened to the yarn? What happened to the saucer?
5. How do you think stone or rust "icicles" form in caves and on the undersides of bridges?
6. What process of the water cycle did you observe in this activity? Explain.
7. How might the process you observed be used to obtain chemicals dissolved in water?
8. What does the water cycle have to do with rocks?

# CUTTING DOWN THE RAIN

Of all Earth's biomes, tropical rain forests probably get the most media coverage. Unfortunately, this is because tropical rain forests are in great danger—they are being cut or burned down at an alarmingly fast rate. Many wonderful animals and plants will disappear with the forests. In addition, the loss of the forests may change the climate of the tropics. How? Find out in this activity.

## What Do I Do?

1. Obtain a medium-sized plant with several branches, two plastic sandwich bags, two twist ties, and a pair of scissors.
2. Select two branches that are as similar as possible in size and number of leaves.
3. Cover one of the branches with a plastic bag. Secure the bag with the twist tie. Be careful not to break or damage the branch while putting on the bag. This branch represents an intact tropical rain forest.
4. Using the scissors, snip off all the leaves on the other branch. Then cover the branch as in step 3. This branch represents an area in which the tropical rain forest has been cut down.
5. Put the plant in a sunny spot. Observe the plant the next day.

## What Did I Learn?

1. Describe what you observed. Where do you think the water came from?
2. What can you infer about the amounts of water put into the air by intact forests and by cleared areas?
3. How do plants fit into the water cycle? (*Hint:* If you have trouble answering this question, review Section 2–2.)
4. Explain why the destruction of forests might lead to change in rainfall.
5. Why should people who live in other biomes be concerned about what is going on in the tropical rain forests?

# GRANDEUR IN THE GRASS

The most noticeable grasslands animals are undoubtedly the large grass-eaters, such as the bison of the North American prairie, kangaroos of the Australian outback, zebras of the African savanna, saiga antelope of the Asian steppe, and rhea (an ostrichlike bird) of the South American pampas. However, grasslands teem with smaller, less noticeable organisms. Take a closer look at grasslands by doing this activity.

## Materials

terrarium case or aquarium with a tightly-fitting fine screen cover
coarse gravel
activated charcoal
sand
potting soil
notebook and pencil
plastic bags and jars with covers for collecting organisms
plants and animals collected from a grassy field

## Procedure

1. Put a layer of gravel 3 cm deep in the terrarium. Add some more gravel to the back of the terrarium to make one or two low hills. Sprinkle a little activated charcoal on the gravel.
2. Mix three parts of potting soil to one part of sand. Spread the sandy soil about 7 cm deep over the gravel, following the hills and valleys in the gravel.
3. Obtain the proper permission to collect organisms from a grassy field. Before you start collecting, look around. What kinds of plants do you see? What kinds of animals? Record your observations in your notebook in the form of drawings and written notes.
4. Collect grasses and a few other small plants from a grassy field. Be sure to include the roots and some of the soil around the roots. Put the plants in a plastic bag to protect them from drying out.
5. Catch a few small animals, such as insects, earthworms, and one or two spiders. Use the jars to hold your animals. **Note:** *Earthworms must be kept cool and moist, so make sure you put damp soil or moist paper towels in the jar with the earthworms.* Make sure that there are air holes in the covers of the jars and that the jars are covered securely.
6. Plant the plants you have collected in the terrarium. Water the plants. Then carefully transfer your animals from the jars into the terrarium.
7. Keep the terrarium in a sunny place. Water the plants every few days with a fine stream of water. Make sure that spiders and other insect-eaters have enough to eat. Observe the terrarium daily and record your observations in your notebook. Remember to record any changes you make in the terrarium, such as watering or adding new insects.

*(continued)*

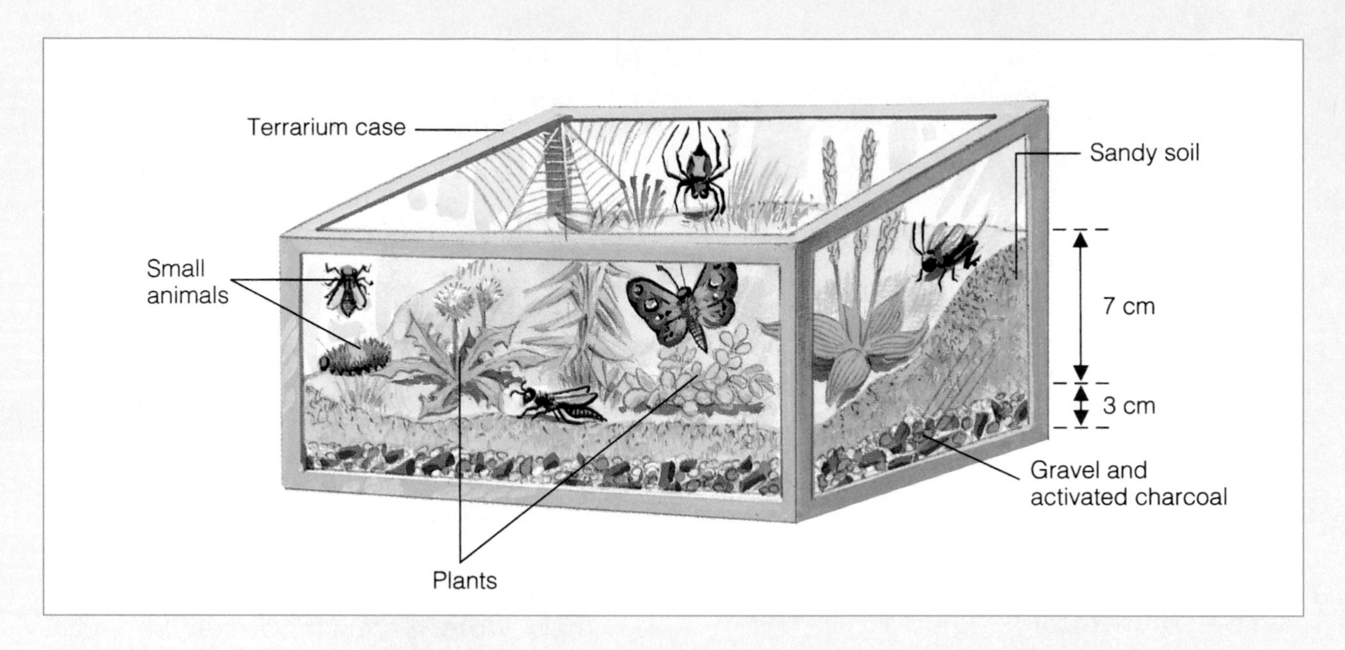


## Observations and Conclusions

1. Make a labeled drawing of your terrarium that shows what kinds of plants and animals it contains.
2. How does your terrarium change over time? Why do you think these changes occur?
3. How do the living and nonliving things in the terrarium interact with one another? How are they dependent on one another?
4. Compare your terrarium with those prepared by other groups in your class. How are they similar? How are they different? How can these similarities and differences be explained?
5. Prepare a poster to share what you have learned about smaller grasslands organisms.

# PAPER ROUTE

A simple, yet effective conservation measure is recycling. In recycling, wastes such as scrap paper, old cans, empty plastic soda bottles, and broken glass jars are used as the raw materials for making new items. For example, old aluminum beverage cans are melted down to produce metal that is used for foil wrap, new beverage cans, and other useful products. In this activity you will try your hand at recycling old newspapers, junk mail, scraps of cloth and thread, and other odds and ends.

## Materials

scrap paper (from sources such as newspapers, magazines, junk mail, notebook paper, and construction paper)
small amount of scraps of fabric and thread (optional)
scissors (optional)
1000-mL (1-L) beaker
hot water
laundry starch
dishpan, about 30 × 34 × 13 cm
egg beater
wood frame, about 25 × 30 cm
4 push pins
piece of window screen, about 25 × 30 cm
rolling pin
blotting paper

## Procedure

1. Tear the paper into pieces less than 5 cm across. If you have fabric, rip or cut it into pieces less than 1 cm across. If you have thread, cut it into pieces less than 5 cm long. Put the pieces in the beaker. When the beaker is full, dump the pieces into the dishpan. Prepare about 1500 mL of pieces.
2. Add 6 L hot water and 360 mL starch to the dishpan.
3. Taking turns, use the eggbeater to beat the mixture in the dishpan until it is about the consistency of pancake batter or white glue. The beaten mixture is called pulp.
4. Attach the screen to the frame with the push pins, as shown in the illustration on page G152.
5. Slide the frame, screen-side up, into the pulp. Wriggle the frame a little so that bits of pulp are distributed evenly across the screen. Lift the screen straight up from the pan and let the water drip into the pan. Repeat this procedure two or three more times, or until the screen is completely covered by a layer of pulp several millimeters thick.
6. Unpin the screen. Carefully put the screen and the wet sheet of pulp on a piece of blotting paper. Cover the screen and sheet of pulp with another piece of blotting paper.
7. Firmly roll the rolling pin over the blotting paper to press the excess water from the pulp sheet.
8. Flip the "sandwich" of blotting paper over. Carefully remove the top piece of blotting paper and the piece of screen to reveal the piece of recycled paper that you have made.
9. Allow the recycled paper to dry. Then peel it off the bottom piece of blotting paper.

*(continued)*

## Observations and Conclusions

1. Describe the paper you made. Why is the paper said to be recycled?
2. Compare your sheet of paper to those made by your classmates. How do they differ? What do you think caused these differences?
3. How is your paper different from commercially made recycled paper? What do you think causes these differences?
4. What are some possible uses for the paper you made?
5. In your own words, explain why recycling is important.

# Appendix A

# THE METRIC SYSTEM

The metric system of measurement is used by scientists throughout the world. It is based on units of ten. Each unit is ten times larger or ten times smaller than the next unit. The most commonly used units of the metric system are given below. After you have finished reading about the metric system, try to put it to use. How tall are you in metrics? What is your mass? What is your normal body temperature in degrees Celsius?

## Commonly Used Metric Units

**Length** The distance from one point to another

| | |
|---|---|
| meter (m) | A meter is slightly longer than a yard.<br>1 meter = 1000 millimeters (mm)<br>1 meter = 100 centimeters (cm)<br>1000 meters = 1 kilometer (km) |

**Volume** The amount of space an object takes up

| | |
|---|---|
| liter (L) | A liter is slightly more than a quart.<br>1 liter = 1000 milliliters (mL) |

**Mass** The amount of matter in an object

| | |
|---|---|
| gram (g) | A gram has a mass equal to about one paper clip.<br>1000 grams = 1 kilogram (kg) |

**Temperature** The measure of hotness or coldness

| | |
|---|---|
| degrees Celsius (°C) | 0°C = freezing point of water<br>100°C = boiling point of water |

## Metric–English Equivalents

2.54 centimeters (cm) = 1 inch (in.)
1 meter (m) = 39.37 inches (in.)
1 kilometer (km) = 0.62 miles (mi)
1 liter (L) = 1.06 quarts (qt)
250 milliliters (mL) = 1 cup (c)
1 kilogram (kg) = 2.2 pounds (lb)
28.3 grams (g) = 1 ounce (oz)
$°C = 5/9 \times (°F - 32)$

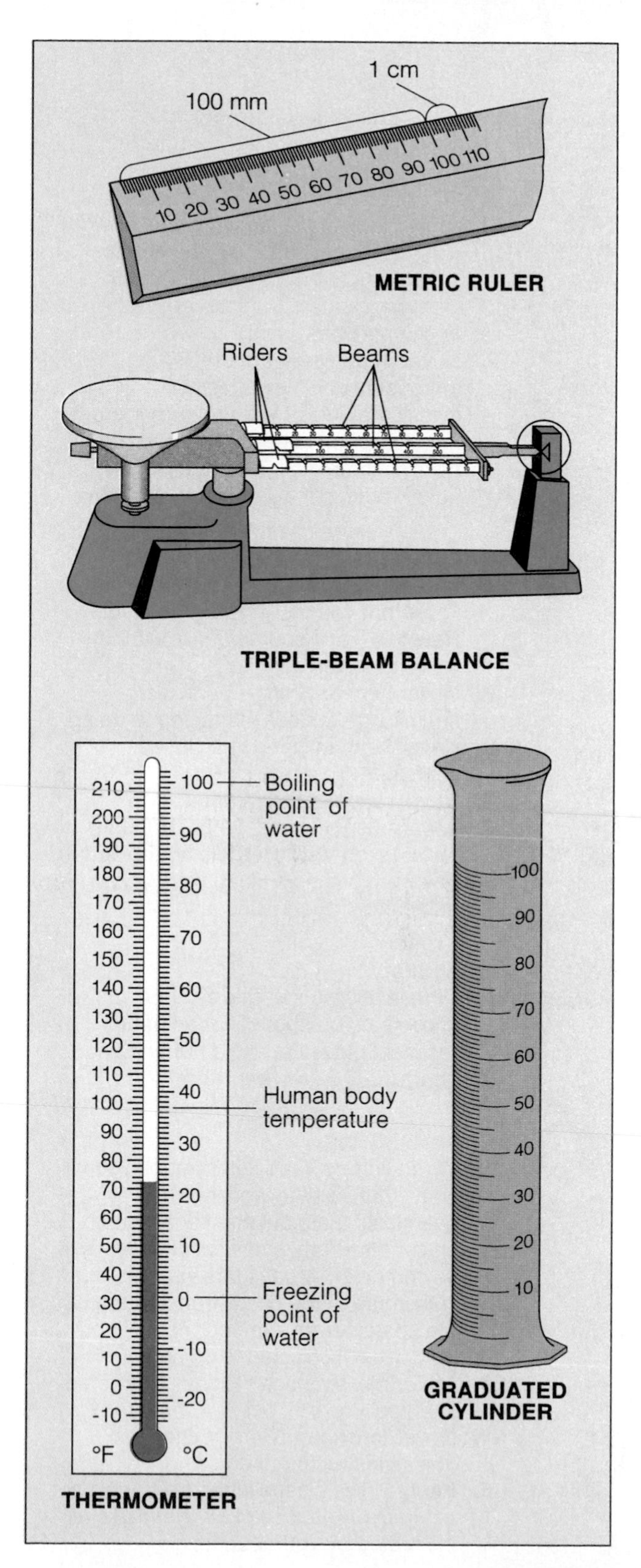

# Appendix B

## LABORATORY SAFETY
### Rules and Symbols

**Glassware Safety**

1. Whenever you see this symbol, you will know that you are working with glassware that can easily be broken. Take particular care to handle such glassware safely. And never use broken or chipped glassware.
2. Never heat glassware that is not thoroughly dry. Never pick up any glassware unless you are sure it is not hot. If it is hot, use heat-resistant gloves.
3. Always clean glassware thoroughly before putting it away.

**Fire Safety**

1. Whenever you see this symbol, you will know that you are working with fire. Never use any source of fire without wearing safety goggles.
2. Never heat anything—particularly chemicals—unless instructed to do so.
3. Never heat anything in a closed container.
4. Never reach across a flame.
5. Always use a clamp, tongs, or heat-resistant gloves to handle hot objects.
6. Always maintain a clean work area, particularly when using a flame.

**Heat Safety**

Whenever you see this symbol, you will know that you should put on heat-resistant gloves to avoid burning your hands.

**Chemical Safety**

1. Whenever you see this symbol, you will know that you are working with chemicals that could be hazardous.
2. Never smell any chemical directly from its container. Always use your hand to waft some of the odors from the top of the container toward your nose—and only when instructed to do so.
3. Never mix chemicals unless instructed to do so.
4. Never touch or taste any chemical unless instructed to do so.
5. Keep all lids closed when chemicals are not in use. Dispose of all chemicals as instructed by your teacher.
6. Immediately rinse with water any chemicals, particularly acids, that get on your skin and clothes. Then notify your teacher.

**Eye and Face Safety**

1. Whenever you see this symbol, you will know that you are performing an experiment in which you must take precautions to protect your eyes and face by wearing safety goggles.
2. When you are heating a test tube or bottle, always point it away from you and others. Chemicals can splash or boil out of a heated test tube.

**Sharp Instrument Safety**

1. Whenever you see this symbol, you will know that you are working with a sharp instrument.
2. Always use single-edged razors; double-edged razors are too dangerous.
3. Handle any sharp instrument with extreme care. Never cut any material toward you; always cut away from you.
4. Immediately notify your teacher if your skin is cut.

**Electrical Safety**

1. Whenever you see this symbol, you will know that you are using electricity in the laboratory.
2. Never use long extension cords to plug in any electrical device. Do not plug too many appliances into one socket or you may overload the socket and cause a fire.
3. Never touch an electrical appliance or outlet with wet hands.

**Animal Safety**

1. Whenever you see this symbol, you will know that you are working with live animals.
2. Do not cause pain, discomfort, or injury to an animal.
3. Follow your teacher's directions when handling animals. Wash your hands thoroughly after handling animals or their cages.

# Appendix C SCIENCE SAFETY RULES

One of the first things a scientist learns is that working in the laboratory can be an exciting experience. But the laboratory can also be quite dangerous if proper safety rules are not followed at all times. To prepare yourself for a safe year in the laboratory, read over the following safety rules. Then read them a second time. Make sure you understand each rule. If you do not, ask your teacher to explain any rules you are unsure of.

## Dress Code

**1.** Many materials in the laboratory can cause eye injury. To protect yourself from possible injury, wear safety goggles whenever you are working with chemicals, burners, or any substance that might get into your eyes. Never wear contact lenses in the laboratory.

**2.** Wear a laboratory apron or coat whenever you are working with chemicals or heated substances.

**3.** Tie back long hair to keep it away from any chemicals, burners and candles, or other laboratory equipment.

**4.** Remove or tie back any article of clothing or jewelry that can hang down and touch chemicals and flames.

## General Safety Rules

**5.** Read all directions for an experiment several times. Follow the directions exactly as they are written. If you are in doubt about any part of the experiment, ask your teacher for assistance.

**6.** Never perform activities that are not authorized by your teacher. Obtain permission before "experimenting" on your own.

**7.** Never handle any equipment unless you have specific permission.

**8.** Take extreme care not to spill any material in the laboratory. If a spill occurs, immediately ask your teacher about the proper cleanup procedure. Never simply pour chemicals or other substances into the sink or trash container.

**9.** Never eat in the laboratory.

**10.** Wash your hands before and after each experiment.

## First Aid

**11.** Immediately report all accidents, no matter how minor, to your teacher.

**12.** Learn what to do in case of specific accidents, such as getting acid in your eyes or on your skin. (Rinse acids from your body with lots of water.)

**13.** Become aware of the location of the first-aid kit. But your teacher should administer any required first aid due to injury. Or your teacher may send you to the school nurse or call a physician.

**14.** Know where and how to report an accident or fire. Find out the location of the fire extinguisher, phone, and fire alarm. Keep a list of important phone numbers—such as the fire department and the school nurse—near the phone. Immediately report any fires to your teacher.

## Heating and Fire Safety

**15.** Again, never use a heat source, such as a candle or burner, without wearing safety goggles.

**16.** Never heat a chemical you are not instructed to heat. A chemical that is harmless when cool may be dangerous when heated.

**17.** Maintain a clean work area and keep all materials away from flames.

**18.** Never reach across a flame.

**19.** Make sure you know how to light a Bunsen burner. (Your teacher will demonstrate the proper procedure for lighting a burner.) If the flame leaps out of a burner toward you, immediately turn the gas off. Do not touch the burner. It may be hot. And never leave a lighted burner unattended!

**20.** When heating a test tube or bottle, always point it away from you and others. Chemicals can splash or boil out of a heated test tube.

**21.** Never heat a liquid in a closed container. The expanding gases produced may blow the container apart, injuring you or others.

**22.** Before picking up a container that has been heated, first hold the back of your hand near it. If you can feel the heat on the back of your hand, the container may be too hot to handle. Use a clamp or tongs when handling hot containers.

## Using Chemicals Safely

**23.** Never mix chemicals for the "fun of it." You might produce a dangerous, possibly explosive substance.

**24.** Never touch, taste, or smell a chemical unless you are instructed by your teacher to do so. Many chemicals are poisonous. If you are instructed to note the fumes in an experiment, gently wave your hand over the opening of a container and direct the fumes toward your nose. Do not inhale the fumes directly from the container.

**25.** Use only those chemicals needed in the activity. Keep all lids closed when a chemical is not being used. Notify your teacher whenever chemicals are spilled.

**26.** Dispose of all chemicals as instructed by your teacher. To avoid contamination, never return chemicals to their original containers.

**27.** Be extra careful when working with acids or bases. Pour such chemicals over the sink, not over your workbench.

**28.** When diluting an acid, pour the acid into water. Never pour water into an acid.

**29.** Immediately rinse with water any acids that get on your skin or clothing. Then notify your teacher of any acid spill.

## Using Glassware Safely

**30.** Never force glass tubing into a rubber stopper. A turning motion and lubricant will be helpful when inserting glass tubing into rubber stoppers or rubber tubing. Your teacher will demonstrate the proper way to insert glass tubing.

**31.** Never heat glassware that is not thoroughly dry. Use a wire screen to protect glassware from any flame.

**32.** Keep in mind that hot glassware will not appear hot. Never pick up glassware without first checking to see if it is hot. See #22.

**33.** If you are instructed to cut glass tubing, fire-polish the ends immediately to remove sharp edges.

**34.** Never use broken or chipped glassware. If glassware breaks, notify your teacher and dispose of the glassware in the proper trash container.

**35.** Never eat or drink from laboratory glassware. Thoroughly clean glassware before putting it away.

## Using Sharp Instruments

**36.** Handle scalpels or razor blades with extreme care. Never cut material toward you; cut away from you.

**37.** Immediately notify your teacher if you cut your skin when working in the laboratory.

## Animal Safety

**38.** No experiments that will cause pain, discomfort, or harm to mammals, birds, reptiles, fishes, and amphibians should be done in the classroom or at home.

**39.** Animals should be handled only if necessary. If an animal is excited or frightened, pregnant, feeding, or with its young, special handling is required.

**40.** Your teacher will instruct you as to how to handle each animal species that may be brought into the classroom.

**41.** Clean your hands thoroughly after handling animals or the cage containing animals.

## End-of-Experiment Rules

**42.** After an experiment has been completed, clean up your work area and return all equipment to its proper place.

**43.** Wash your hands after every experiment.

**44.** Turn off all burners before leaving the laboratory. Check that the gas line leading to the burner is off as well.

# Glossary

## Pronunciation Key

When difficult names or terms first appear in the text, they are respelled to aid pronunciation. A syllable in SMALL CAPITAL LETTERS receives the most stress. The key below lists the letters used for respelling. It includes examples of words using each sound and shows how the words would be respelled.

| Symbol | Example | Respelling |
|---|---|---|
| a | hat | (hat) |
| ay | pay, late | (pay), (layt) |
| ah | star, hot | (stahr), (haht) |
| ai | air, dare | (air), (dair) |
| aw | law, all | (law), (awl) |
| eh | met | (meht) |
| ee | bee, eat | (bee), (eet) |
| er | learn, sir, fur | (lern), (ser), (fer) |
| ih | fit | (fiht) |
| igh | mile, sigh | (mighl), (sigh) |
| oh | no | (noh) |
| oi | soil, boy | (soil), (boi) |
| oo | root, rule | (root), (rool) |
| or | born, door | (born), (dor) |
| ow | plow, out | (plow), (owt) |

| Symbol | Example | Respelling |
|---|---|---|
| u | put, book | (put), (buk) |
| uh | fun | (fuhn) |
| yoo | few, use | (fyoo), (yooz) |
| ch | chill, reach | (chihl), (reech) |
| g | go, dig | (goh), (dihg) |
| j | jet, gently, bridge | (jeht), (JEHNT-lee), (brihj) |
| k | kite, cup | (kight), (kuhp) |
| ks | mix | (mihks) |
| kw | quick | (kwihk) |
| ng | bring | (brihng) |
| s | say, cent | (say), (sehnt) |
| sh | she, crash | (shee), (krash) |
| th | three | (three) |
| y | yet, onion | (yeht), (UHN-yuhn) |
| z | zip, always | (zihp), (AWL-wayz) |
| zh | treasure | (TREH-zher) |

**biogeography:** the study of where plants and animals live throughout the world

**biological clock:** an internal timer that keeps track of a cycle of time and helps an organism stay in step with rhythmic cycles of change in the environment

**biome:** a division based on climate, plants, and animals; an environment that has a characteristic type of climax community

**canopy:** the layer of a forest biome that consists of the tops of trees; the "roof" of a forest

**captive breeding:** the practice of getting animals in zoos to have offspring

**carbon cycle:** the cyclical series of processes in which carbon moves through the living and nonliving parts of the environment

**climax community:** the stable community that is the final stage of succession

**commensalism:** a form of symbiosis in which one organism benefits and the other is not harmed

**community:** the living part of an ecosystem

**competition:** the type of interaction in which organisms struggle with one another to obtain resources

**conifer**: a plant, usually an evergreen tree, that produces its seeds in cones

**consumer:** an organism that cannot make its own food

**decomposer:** an organism that breaks down the bodies of dead organisms into simpler substances

**deforestation:** the destruction of forests

**desertification** (dih-zert-uh-fih-KAY-shuhn): the process in which desertlike conditions are created where there had been none in the recent past
**dispersal:** the movement of living things from one place to others; spreading out
**diurnal** (digh-ER-nuhl): active during the day
**ecological succession:** the process in which the community in a particular place is gradually replaced by another community
**ecology:** the study of the relationships and interactions of living things with one another and with their environment
**ecosystem:** a unit consisting of all the living and nonliving things in a given area that interact with one another
**endangered:** in danger of becoming extinct
**environment:** all the living and nonliving things with which an organism may interact
**estivation:** a summer resting state
**estuary** (EHS-tyoo-air-ee): an environment found at the boundary between fresh water and salt water that contains a mixture of fresh water and salt water
**exotic species:** a species that is not native to a place
**extinct:** no longer in existence; used to describe subspecies, species, and so on, in which there are no living individuals
**food chain:** a representation of a series of events in which food energy and matter are transferred from one organism to another
**food web:** a diagram that consists of many overlapping food chains
**freshwater biome:** the biome that consists of the Earth's bodies of fresh water, such as lakes, ponds, streams, and rivers
**habitat:** the place in which an organism lives and obtains the resources it needs to survive
**hibernation:** a winter resting state
**host:** an organism that provides a home for another organism; in parasitism, the organism that is harmed by the parasite
**marine biome:** the ocean biome
**migration:** the movement of organisms from one place to another in response to periodic environmental changes; usually refers to cyclical movements
**mutualism:** a form of symbiosis in which both organisms benefit
**niche** (NIHCH): an organism's role in an ecosystem, which includes everything the organism does and everything the organism needs in its environment
**nitrogen cycle:** the cyclical series of processes in which nitrogen moves from the nonliving parts of the environment to living things and back again
**nocturnal** (nahk-TER-nuhl): active during the night
**oxygen cycle:** the cyclical series of processes in which oxygen moves through the living and nonliving parts of the environment
**parasite:** an organism that lives on or inside the body of a host organism and harms the host
**parasitism:** a form of symbiosis in which one organism benefits and the other is harmed
**permafrost:** the layer of permanently frozen soil in the tundra
**phytoplankton:** microscopic producers (organisms that can make their own food) that live near the surface of the ocean and other bodies of water
**population:** a group of organisms of the same species living together in the same area
**predator:** an organism that kills and eats another organism
**prey:** an organism that is eaten by a predator
**producer:** an organism that is able to make its own food by using a source of energy to turn simple raw materials into food
**symbiosis** (sihm-bigh-OH-sihs; plural: symbioses): a close relationship between two organisms in which one organism lives near, on, or even inside another organism and in which at least one organism benefits
**taiga:** the northernmost coniferous forest biome
**water cycle:** the cyclical series of processes in which water moves through the living and nonliving parts of the environment
**wildlife conservation:** the intelligent management of living resources so that they provide the greatest possible benefit for the longest possible time

# Index

Credits

**Cover Background:** Ken Karp
**Photo Research:** Natalie Goldstein
**Contributing Artists:** Illustrations: Warren Budd & Associates, Ltd.; Wende Caporale/Gwen Goldstein, Art Representatives; Raymond Smith. Charts and graphs: Function Thru Form
**Photographs:** **4** Wayne Lynch/DRK Photo; **5** top: Michael Fogden/DRK Photo; bottom: C. H. Robinson/Animals Animals/Earth Scenes; **6** top: Lefever/Grushow/Grant Heilman Photography; bottom: Index Stock; center: Rex Joseph; **8** top left: C. H. Robinson/Animals Animals/Earth Scenes; bottom left: Julia Sims/Peter Arnold, Inc.; right: Larry Ulrich/DRK Photo; **9** Tom Bledsoe/DRK Photo; **10** and **11** Milton Rand/Tom Stack & Associates; **12** top left: P. David/Planet Earth Pictures, bottom left: Michael Fogden/Animals Animals/Earth Scenes; right: Kjell B. Sandved; **13** top: Zig Leszczynski/Animals Animals/Earth Scenes; bottom: Kjell B. Sandved; **14** D. Cavagnaro/DRK Photo; **15** top: Scott Blackman/Tom Stack & Associates; bottom: Breck P. Kent; **16** top and bottom: Frans Lanting/Minden Pictures, Inc.; center: John Cancalosi/DRK Photo; **18** top: Breck P. Kent; bottom: John Shaw/Tom Stack & Associates; **19** top left: Terry Domico/Earth Images; top right: Tim Davis/Photo Researchers, Inc.; bottom: Jeffrey L. Rotman; **20** S. Nielsen/Imagery; **21** top and center: T. A. Wiewandt/DRK Photo; bottom: John Cancalosi/DRK Photo; **24** left: Hans and Judy Beste/Animals Animals/Earth Scenes; right: F. Gohier/Photo Researchers, Inc.; **25** Stouffer Productions/Animals Animals/Earth Scenes; **26** top to bottom: Frans Lanting/Minden Pictures, Inc.; S. Nielsen/DRK Photo; Wilbur Samuel Tripp; **27** border and right center: Scott Camazine/Photo Researchers, Inc.; top and left center: Timothy Eagan/Woodfin Camp & Associates; bottom right: Lilia I. deGuzman/USDA; **28** top: Lee Lyon/Survival Anglia; bottom: Peter Ward/Bruce Coleman, Inc.; **29** left: Charles Seaborn/Woodfin Camp & Associates; right: Stephen J. Krasemann/DRK Photo; **30** top left: Dwight Kuhn Photography; bottom left: Ashod Francis/Animals Animals/Earth Scenes; right: Stephen J. Krasemann/DRK Photo; **31** left: Stephen Dalton/Natural History Photographic Agency; top right: Michael Fogden/DRK Photo; bottom right: Gary Milburn/Tom Stack & Associates; **32** top: Jim Brandenburg/Minden Pictures, Inc.; bottom: Wayne Lynch/DRK Photo; **33** top left and top center: Michael Fogden/DRK Photo; top right: Stephen J. Krasemann/DRK Photo; center right: David Denning/Earth Images; bottom right: T. A. Wiewandt/DRK Photo; **34** left: Patti Murray/Animals Animals/Earth Scenes; center and right: Michael Fogden/DRK Photo; **35** top: M.P. Kahl/DRK Photo; bottom: Stephen J. Krasemann/DRK Photo; **36** top left: Superstock; top right: Kevin Schafer/Tom Stack & Associates; center: James Mason/Black Star; bottom: Wolfgang Kaehler; **37** top: J. Langevin/Sygma; bottom left: Craig Aurness/Woodfin Camp & Associates; bottom right: John Gerlach/Animals Animals/Earth Scenes; **39** Dwight Kuhn Photography; **40** Ken Karp; **43** Gary W. Griffen/Animals Animals/Earth Scenes; **44** and **45** Jeff Foott/DRK Photo; **46** top and right: T. A. Wiewandt/Natural History Photography; bottom left: Frans Lanting/Minden Pictures, Inc.; **47** left: E.R. Degginger/Picture Perfect; right: Kjell B. Sandved; **48** Paul Fusco/Magnum Photos, Inc.; **49** left: Dwight Kuhn Photography; right, top and bottom: Gary Milburn/Tom Stack & Associates; **50** top left: F. Stuart Westmorland/Tom Stack & Associates; top right: Joe McDonald/Animals Animals/Earth Scenes; bottom left: Zig Leszczynski/Breck P. Kent; bottom right: Robert & Linda Mitchell Photography; **51** E. R. Degginger/Animals Animals/Earth Scenes; **52** left: Roger Garwood/Colorific; right: Johnny Johnson/DRK Photo; **53** left: Frans Lanting/Minden Pictures, Inc.; top right: Breck P. Kent; bottom right: T. A. Wiewandt/Natural History Photography; **54** Lick Observatory, University of California; **55** left: Breck P. Kent; top right: NASA; bottom right: Robert & Linda Mitchell Photography; **57** Pat Crowe/Animals Animals/Earth Scenes; **59** top left: Dan McCoy/R. Langridge/Rainbow; top right: Dwight Kuhn Photography; bottom: DPI; **60** left: Doug Wechsler/Animals Animals/Earth Scenes; right: Dwight Kuhn Photography; **61** top to bottom: Breck P. Kent; **62** Jeff Foott Productions; **63** top and bottom: Gary Milburn/Tom Stack & Associates; **67** Vince Streano/Stock Market; **68** and **69** Wolfgang Kaehler; **70** top: David Macdonald/Oxford Scientific Films/ Animals Animals/Earth Scenes; bottom left: Frans Lanting/Minden Pictures, Inc.; bottom right: Nancy Adams; **71** top: Jack Swenson/Tom Stack & Associates; bottom left: Ed Degginger/Animals Animals/Earth Scenes; bottom right: Zig Leszczynski/Animals Animals/Earth Scenes; **72** Frans Lanting/Minden Pictures, Inc.; **73** top: Breck P. Kent; bottom: Jeff Foott Productions; **75** top and bottom left: Stephen J. Krasemann/DRK Photo; bottom right: Wolfgang Kaehler; **76** top: Chase Swift/Tom Stack & Associates; bottom: Stephen J. Krasemann/DRK Photo; **77** top: Wolfgang Kaehler; bottom: S. Nielsen/DRK Photo; **78** top: Leonard Lee Rue III/DPI; bottom left: Joseph R. Pearce/DRK Photo; bottom right: Stephen J. Krasemann/DRK Photo; **79** top left: Jeff Foott/DRK Photo; center: Wayne Lankinen/DRK Photo; top right and bottom: Stephen J. Krasemann/DRK Photo; right center: Robert Frerck/Odyssey Productions; **80** left: Zig Leszczynski/Animals Animals/Earth Scenes; right: Stephen J. Krasemann/DRK Photo; **81** top: Michael Fogden/DRK Photo; bottom: Breck P. Kent; **82** top left and right: Michael Fogden/DRK Photo; bottom: Wolfgang Kaehler; **84** top left: Frans Lanting/Minden Pictures, Inc.; top center: Kjell B. Sandved; top right: T. A. Wiewandt/Natural History Photo; bottom: Wolfgang Kaehler; **85** left: Wolfgang Kaehler; right: Breck P. Kent; **86** left: Patti Murray/Animals Animals/Earth Scenes; right: Robert Freck/Odyssey Productions; **87** left: Frans Lanting/Minden Pictures, Inc.; right: Anthony Bannister/Animals Animals/Earth Scenes; bottom: Wolfgang Kaehler; **88** top: Breck P. Kent; bottom left: Lewis Trusty/Animals Animals/Earth Scenes; bottom right: Jeff Foott Productions; **89** top left: Paul Humann/Jeffrey L. Rotman; top right: Doug Perrine/DRK Photo; center: Frans Lanting/Minden Pictures, Inc.; bottom: Peter David/Planet Earth Pictures; **90** top: Dwight Kuhn Photography; center: Johnny Johnson/DRK Photo; bottom left: Robert P. Comport/Animals Animals/Earth Scenes; bottom right: Zig Leszczynski/Animals Animals/Earth Scenes; **91** left: Breck P. Kent; right: Dan McCoy/Rainbow; **92** left: Fred Whitehead/Animals Animals/Earth Scenes; right: Timothy A. Murphy/Image Bank; **93** left: T. A. Wiewandt/Natural History Photo; right: Ed Degginger/Animals Animals/Earth Scenes; **97** Breck P. Kent; **98** and **99** Stephen J. Krasemann/DRK Photo; **100** top: © National Museums of Scotland; bottom: Illustration by John Tenniel from *Alice in Wonderland and Through the Looking Glass* by Lewis Carroll; **101** Stephen J. Krasemann/DRK Photo; **102** top left and bottom right: Frans Lanting/Minden Pictures, Inc.; top right: Robert & Linda Mitchell Photography; bottom left: Stephen J. Krasemann/DRK Photo; **103** left: Don & Pat Valenti/DRK Photo; top right: John Stern/Animals Animals/Earth Scenes; bottom right: Ron Kimball; **104** top: Doug Wechsler/Animals Animals/Earth Scenes; bottom: Dr. Nigel Smith/Animals Animals/Earth Scenes; **105** top left: Dan McCoy/Rainbow; top right: Wolfgang Kaehler; bottom left: Tom Bean/DRK Photo; bottom right: Frans Lanting/Minden Pictures, Inc.; **106** top left and bottom right: Kevin Schafer/Tom Stack & Associates; top right: Michael Fogden/DRK Photo; center left: Gary Milburn/Tom Stack & Associates; bottom left: Richard Kolar/Animals Animals/Earth Scenes; **107** top: Alain Keller/Sygma; bottom: Richard Hoffmann/Sygma; **108** left: Stephen J. Krasemann/DRK Photo; top: Tom Bean; center: Dan McCoy/Rainbow; right: Patricia Caulfield/Animals Animals/Earth Scenes; **109** Frans Lanting/Minden Pictures, Inc.; **110** top left: Fred Whitehead/Animals Animals/Earth Scenes; top right: Gordon Rodda/Photo Resource Hawaii; bottom: Johnny Johnson/Animals Animals/Earth Scenes; **111** top left: Zigy Kaluzny; top center: Frans Lanting/Minden Pictures, Inc.; top right: Jeff Foott Productions; bottom: Wolfgang Kaehler; **112** top: John Cancalosi/DRK Photo; bottom: Frans Lanting/Minden Pictures, Inc.; **113** top: Larry Lefever/Grant Heilman Photography; bottom: Robert Frerck/Odyssey Productions; **114** Annie Griffiths/DRK Photo; **115** Chase Swift/Tom Stack & Associates; **116** Lowell Georgia/Photo Researchers, Inc.; **117** top left: Tom & Pat Leeson/DRK Photo; top right: Kjell B. Sandved; center: Mark Boulton/Photo Researchers, Inc.; bottom: San Francisco Zoo; **118** left and right: David Muench Photography Inc.; center: Robert C. Simpson/Tom Stack & Associates; **119** left: Cincinnati Zoo; right: Ron Garrison/Zoological Society of San Diego; **120** top: Ron Garrison/Zoological Society of San Diego; center: Belinda Wright/DRK Photo; bottom: John Chellman/Animals Animals/Earth Scenes; **121** top left: Gary Milburn/Tom Stack & Associates; top right: Robert & Linda Mitchell Photography; bottom: J. Robinson/Animals Animals/Earth Scenes; **122** left: Baker/Greenpeace; right: Rick Falco/Sipa Press; **127** P. La Tourrette/VIREO/Academy of Natural Sciences, Philadelphia; **128** and **129** Peter Menzel; **132** Tom McHugh/Photo Researchers, Inc.; **135** Stephen J. Krasemann/Photo Researchers, Inc.; **136** left: T.A. Wiewandt/Natural History Photography; right: S. Nielsen/Imagery; **157** NASA